EARTHQUAKE FORECASTING AND WARNING

DEVELOPMENTS IN EARTH AND PLANETARY SCIENCES

EARTHQUAKE FORECASTING AND WARNING

Tsuneji Rikitake

Developments in Earth and Planetary Sciences

03

Center for Academic Publications Japan/Tokyo
D. Reidel Publishing Company/Dordrecht · Boston · London

Library of Congress Cataloging in Publication Data

DATA APPEARS ON SEPARATE CARD

ISBN 90-277-1218-2

Published by Center for Academic Publications Japan, Tokyo, in co-publication with D. Reidel Publishing Company, P. O. Box 17, 3300 AA Dordrecht.

Sold and distributed in Japan, China, Korea, Taiwan, Indonesia, Cambodia, Laos, Malaysia, Philippines, Thailand, Vietnam, Burma, Pakistan, India, Bangla Desh, Sri Lanka by Center for Academic Publications Japan, 4–16, Yayoi 2-chome, Bunkyo-ku, Tokyo 113, Japan.

Sold and distributed in the U.S.A. and Canada by Kluwer Boston Inc., 190 Old Derby Street, Hingham, MA 02043, U.S.A.

Sold and distributed in all other countries by Kluwer Academic Publishers Group, P. O. Box 322, 3300 AH Dordrecht, Holland.

Printed in Japan

Damage to the library building of the Tangshan Mining College by the 1976 Tangshan earthquake of magnitude 7.8 (photo by Rikitake, 1981).

Damage to the steam locomotive engine factory by the 1976 Tangshan earthquake of magnitude 7.8 (photo by Rikitake, 1981).

CONTENTS

PREFACE

Studies related to the earth and planets along with their surroundings are of great concern for modern scientists. Global geodynamics as represented by plate tectonics has now become one of the most powerful tools by which we can study the causes of earthquakes, volcanic eruptions, mountain formation and the like. Various missions sent out to space, manned or unmanned, brought out geoscientific features of the moon, Mars, Venus and other planets. Earthquake prediction that was the business of astrologers and fortune-tellers some twenty years ago, has now grown up to be an important science. A number of destructive earthquakes were successfully forecast in the People's Republic of China.

In the light of the above-mentioned and other accomplishments in geosciences, we feel that it is a good thing to publish a series of monographs which review selected topics of earth and planetary sciences. We are of course well aware of the fact that similar monographs have been and will be published from overseas publishers. The series, which we plan to publish, will therefore stress Japanese work. But we hope that the series will also include review articles by distinguished overseas authors.

The series, which is named the "Developments in Earth and Planetary Sciences" will be published by the Center for Academic Publications Japan and the D. Reidel Publishing Company. It is my great pleasure to work as the Editor of the series. I should like to have comments on what subjects we shall choose in future publications. I shall be greatly obliged if anyone would suggest suitable subjects and potential authors for the series to me.

Tsuneji Rikitake
Editor

PREFACE TO VOLUME 3
AND INTRODUCTORY REMARKS

A Symposium on Earthquake Prediction was held at the Headquarters of the United Nations Educational, Scientific, and Cultural Organization (UNESCO) in Paris on April 2–6, 1979. Some 240 scientists participated in the symposium from 40 different countries. The most outstanding and important feature of the symposium consisted of researchers working in the fields of natural and social sciences sharing the discussion. Even administrators responsible for disaster prevention and civil defense and representatives of insurance companies attended the sessions.

It appears to the author that the fact that such an international symposium tackling earthquake forecasting and warning from both sides of natural and social sciences was successfully held indicates that earthquake prediction is now becoming a matter of actuality. It is true that technology of earthquake prediction is still in a developing stage. But social demands for practicable prediction of destructive earthquakes are so strong that every effort should be concentrated toward actual prediction even if the probability of false prediction is not inconsiderable.

The 1975 Haicheng earthquake of magnitude (M) 7.3 in Liaoning Province, China was successfully predicted. In this case, long-, medium-, and short-term predictions were issued along with an imminent prediction several hours prior to the main shock. This was the first prediction of destructive earthquake in history. Three more earthquakes larger than $M=7$ were predicted in China in 1976 although no imminent prediction was issued for the 1976 Tangshan earthquake ($M=7.8$) which killed a few hundred thousand people.

Earthquake prediction programs that started in the 1960's in China, Japan, the U.S.A., and the U.S.S.R. made such remarkable progress that many an earthquake precursor has been identified through intensive geophysical and geochemical observations. Particularly, the development of earthquake prediction research in recent years has been amazingly rapid, so that considerable data pertinent to earthquake prediction are now available.

The author wrote a book entitled "Earthquake Prediction" which was published in 1976 (RIKITAKE, 1976a). This work covered then-available aspects of earthquake prediction. As was anticipated even at that time, the progress of earthquake prediction study has so escalated that substantial new study material can be added to that original work. For instance, 282 examples of earthquake precursor were analysed in that book. According to the author's survey, the number of reported precursors has increased so rapidly since 1976 that 391 geophysical and geochemical precursors were called to his attention by the beginning of 1978.

Extensive networks of seismographs, tiltmeters, strainmeters, tide-gauges, gravimeters, magnetometers, resistivity variometers, radon content monitors, and the like, have been set up in recent years over earthquake-threatened areas in China, Japan, the U.S.A., and the U.S.S.R. It may be said, therefore, that a dramatic intensification of observation systems aimed at actual earthquake prediction has been completed, although further intensification is certainly necessary.

Another aspect of earthquake prediction and warning as advanced in the last 5 years or so is the study of public reaction against a warning of a highly destructive earthquake. It has been pointed out by pilot studies that an untimely issue of earthquake warning could sometimes result in severe social and economic damage which may exceed that caused by the earthquake itself. It is now recognized that the socioeconomic aspects of earthquake prediction and warning should be extensively studied prior to an actual warning.

Special legislations for mitigating earthquake hazards were made in the U.S.A. and Japan respectively in 1977 and 1978. The Japanese law called the "Large-scale Earthquake Countermeasures Act" is unique in the fact that countermeasures against earthquake hazards will be taken on the basis of long- and short-term earthquake prediction information. It may be that the feasibility of achieving actual earthquake prediction has more or less become a nation-wide consensus, at least in Japan, although earthquake prediction is still in a developing stage.

In the light of recent achievements as mentioned above, the author considered it prudent to publish a book that summarizes new developments of earthquake prediction and warning. Actually, a book in Japanese for that purpose has already been written by him and was published by the Center for Academic Publications Japan / Japan Scientific Societies Press in 1979. The author holds the opinion, however, that such a book should also be published in an internationally accessible language such as English, otherwise the book is of little value internationally.

Although Japan is one of the leading countries putting emphasis on earthquake prediction, most data compiled in Japan has been published in Japanese journals in the Japanese language difficult to read for overseas colleagues. Such data, of great importance for earthquake prediction study, should be presented to overseas researchers in an accessible form.

For those reasons the author undertakes anew the writing of the present volume. To tell the truth, however, this book is not a mere translation of the book in Japanese that has already been published. Further advances in the study of earthquake prediction and warning took place even during the relatively short period between the time when the Japanese edition was published and the present. It is planned to make the present book as up-to-date as possible by covering all these new findings.

The author will try in this work not to repeat that written in the previous publication "Earthquake Prediction" except for minimum overlaps that are absolutely necessary for reasoning, so that those who are interested in overall aspects of earthquake prediction are kindly asked to refer to both books.

It was the author's privilege to visit China and see some of the Chinese efforts toward earthquake prediction in September, 1978. He and his delegation were privileged to visit the Songpan-Pingwu area, the epicentral area of the 1976 large earthquakes ($M = 7.2, 6.7, 7.2$), in Sichuan Province, where they were the first foreigners to enter that area. Chinese experiences in relation to forerunners of an earthquake are sometimes so fantastic that it is really hard to grasp the facts. Nevertheless it should be borne in mind that the Chinese succeeded in issuing imminent earthquake predictions 4 times among 5 destructive earthquakes during 1975–76. It seems that macroscopic phenomena such as anomalous animal behavior, fire-ball appearance and the like are sometimes observed very clearly in China forerunning a large earthquake. Although the author does not presume his complete understanding of the Chinese way of earthquake prediction even after two more visits to China in 1980 and 1981, he will attempt in this book to introduce the Chinese approach to earthquake prediction to occidental colleagues as accurately as possible. That the Japanese can read many Chinese characters is an advantage for such a purpose.

This volume consists of two parts. Part I is devoted to summarizing the natural science side of earthquake prediction based on the most recent information. Meanwhile, social consequences of earthquake prediction and warning along with relevant legislations will be dealt with in Part II. Although a geophysicist, the author attempts to cover the topics in Part II, because those subjects are vitally important even to those working on the natural science side of earthquake prediction. It will be the author's pleasure if a sociologist would undertake to further expand his findings.

Part I consists of 7 chapters. In Chapter 1, the author will attempt to summarize the current state of earthquake prediction programs in Japan, the U.S.A., the U.S.S.R., and the People's Republic of China. Dramatic intensification of the Japanese program in recent years can well be recognized by comparing the material contained in Chapter 1 to that stated concerning the Japanese program in the author's previous book. The system of deep-well seismic observation around the Tokyo area, as well as that of sea bottom seismographs as extended off the Pacific coast in the Tokai (literally means East Sea) area, an area between Tokyo and Nagoya under menace of a great earthquake, are quite unique among earthquake prediction observations in the world. Being extended more

than 100 km off the coast and telemetered to the Japan Meteorological Agency (JMA) in Tokyo on real-time basis, the latter
system serves for short-term and imminent prediction of the feared
Tokai earthquake, the magnitude of which is estimated as 8 or so,
should it occur.

An extensive network of observation by volume strainmeters,
tiltmeters, seismographs, tide-gauges, underground-water monitors,
and the like have been spread all over the Tokai area and most of
them are telemetered to JMA on an on-line real-time basis.

A nationwide geodetic survey has been intensified in recent
years. As to microearthquake observations, extensive telementering
systems are now available. Chapter 1 includes a description of the
above developments in the Japanese program in detail along with its
budgetary developments.

The U.S. national project on earthquake prediction launched
in 1973 as a part of the Earthquake Hazards Reduction Program
(EHRP) was mentioned in the previous text. The budget for EHRP
increased discontinuously in 1978.

A remarkable installation is that of the U.S. Geological Survey
(USGS) Californian system of microearthquake observation in
cooperation with the California Institute of Technology (CIT), the
number of seismograph stations exceeding 500. Observations of
various disciplines are also provided along the San Andreas fault.
Special attention has been paid to the anomalous land uplift in
southern California since 1976.

All these U.S. efforts will be reviewed in Chapter 1 on the basis of
published information as well as the author's material gathered from
American colleagues.

Chapter 1 also covers the advances of the U.S.S.R. program on
earthquake prediction although only relatively limited information
has reached the author perhaps because of linguistic difficulty.

In contrast to the very brief description of the Chinese program
on earthquake prediction in the previous text, a fairly detailed report
of the Chinese program is now contained in Chapter 1. A number of
American, Canadian, and Japanese seismological delegations have
visited China in recent years. In return, a few Chinese missions on
seismology visited these countries. It was therefore possible to become

aware of the Chinese program on earthquake prediction through the reports of these exchange programs. It was most fortunate for the author that he and his delegation could visit Peking (Beijing) and its surroundings, Shenyang-Haicheng-Yingkou area in Liaoning Province and Chengdu and Songpan-Pingwu area in Sichuan Province in 1978. The author enjoyed two more visits to China in 1980 and 1981.

That the Chinese constructed an incredibly large number of observation stations with countless workers, that observations by amateurs play an important role on earthquake prediction, and that macroscopic precursors such as anomalous animal behavior (which can be observed by the eye and ear, often observed very clearly in China) will be described in Chapter 1 in relation to the Chinese program on earthquake prediction. It is really interesting that ways and means for actualization of earthquake prediction, sometimes difficult to understand on the basis of western logic, have been adopted with success in China.

Chapters 2 and 3 will be reserved for analysis of earthquake records in history and monitoring of crustal strain by repeating geodetic surveys. They are useful for extremely long-range earthquake forecasting. The importance of historic documentation of earthquakes covering about 3,000 and 2,000 years respectively in China and Japan in relation to monitoring future seismicity has long been recognized although much of existing documentation was only qualitative. In Chapter 2, however, a quantitative study, namely examples of probability estimate for recurrence of great earthquakes at subduction zones, will be demonstrated only from analysis of historical records. Migration tendency of seismicity will also be presented including occurrences of large Sichuan and Yunnan earthquakes in China alternating one with the other. Recent knowledge about seismic gaps will also be one of the topics of this chapter.

Estimation of crustal strain accumulation and statistics of ultimate crustal strain as presented in the previous text have been improved as will be seen in Chapter 3. Discussion is focused at the Tokai area in Central Japan where we have every reason to expect a great earthquake to occur sooner or later.

What will be presented in Chapters 2 and 3 are only concerned with extremely long-range forecasting of earthquake occurrence or, in other words, it is only appropriate to call them the "regionalization." However, much more concrete symptoms forerunning an earthquake must be detected in order to make an actual earthquake prediction. Chapter 4 will be devoted to presenting many recently-reported precursory effects which may foretell occurrence of an earthquake. It is really remarkable that the number of reports on earthquake precursors has increased so enormously in recent years perhaps because of the completion of observation networks under earthquake prediction programs in each country.

It seems likely that there are precursors of various disciplines, i.e. land deformation, change in sea-level, change in tilting of the ground and strain and stress in the earth's crust, foreshock, anomalous seismic activity prior to the main shock, seismic gap of the 2nd kind (decrease in the activity of small earthquakes prior to the main shock), b value (a parameter representing the ratio of number of occurrence of relatively large shocks to that of relatively small shocks), change in source mechanism, hypocentral migration, change in seismic wave velocities, change in earth-tidal amplitude, change in the geomagnetic field intensity, change in the amplitude of short-period geomagnetic variations originating primarily from outside the earth, change in earth-currents, change in earth resistivity, change in gravity, change in undergound water level and chemical content especially that in radon content and so on. It will be planned in Chapter 4 to make the characteristics of these precursors clear on the basis of the now-available data examples amounting to 391 in number.

It is interesting to note that many macroscopic precursors, as exampled by Chinese scientists have been disclosed in recent years on the occasions of strong earthquakes in various countries. They are just observed by the eye, ear, and nose without making use of more sophisticated instruments. The author will try to summarize in Chapter 4 what he learned about anomalous animal behavior, fire-ball appearance, well-water change, and the like forerunning an earthquake as gathered from Chinese colleagues during his visit to China. A few studies on anomalous animal behavior in the U.S.A. as well as in Japan prior to a large earthquake will also be included in this chapter.

Strategies of earthquake prediction currently under consideration in Japan, China, and possibly in other countries will be the subject of Chapter 5. First of all, the rating of earthquake-threatened areas by the Coordinating Committee for Earthquake Prediction (CCEP) will be mentioned. The CCEP, the headquarters of earthquake prediction in Japan, is working hard on detecting premonitory effects over the areas of intensified observation. Starting from an extremely long-term prediction as will be discussed in Chapters 2 and 3, earthquake percursors are to be detected by intensified observations. The occurrence time of an earthquake will then be estimated on the basis of the characteristics of precursors observed. In the case of an actual prediction, the importance of overall judgment based on many precursors will be stressed.

In order to achieve earthquake prediction by means of the strategy as presented in Chapter 5, a system of earthquake prediction must be developed. In Chapter 6 will be described the earthquake prediction systems in Japan, the U.S.A., and China in fair detail. Special mention will be made of the Prediction Council for the Tokai area, Japan, which is under menace of a great earthquake. The Council has been officially called the Prediction Council for the Area under Intensified Measures against Earthquake Disaster as from August 7, 1979 in association with the designation of the Tokai and neighboring areas to the Area under Intensified Measures against Earthquake Disaster under the Large-scale Earthquake Countermeasures Act enacted in 1978 (RIKITAKE, 1979c).

In Chapter 7, the author will attempt to present a comparative study of program, budget, strategy, system, legislation, and the like relevant to earthquake prediction in Japan, the U.S.A., China, and the U.S.S.R. Such a study will be useful for gathering information about the efforts toward earthquake prediction in each country.

Part II, consists of 6 chapters, and will manifest various problems related to conversion of earthquake prediction information into an earthquake warning.

In Chapter 8, case histories of earthquake warning in relation to the Matsushiro, Haicheng, Tangshan, Songpan-Pingwu, and Izu-Oshima Kinkai earthquakes will be presented. It is the author's belief that these examples are of prime importance to provide the basis for studying the various problems concerned.

Possible classification of earthquake prediction information will be discussed in Chapter 9. The author's idea that the prediction information should be classified according to the time-window of possible earthquake occurrence and the reliability of information will be presented in this chapter.

What form would an earthquake warning take? Except for Chinese instances, no actual issuance of earthquake warning of a great earthquake has been experienced, especially in capitalistic countries. In Chapter 10, possible forms of earthquake warning will be discussed. Special mention will be made of the "earthquake warnings statement" to be issued by the Japanese Prime Minister based on the Large-scale Earthquake Countermeasures Act in case an imminent prediction of a large-scale earthquake is made by the Prediction Council.

One of the difficult problems of earthquake warning is to communicate correct information to the public at large. What happened in the case of Izu-Oshima Kinkai earthquake ($M = 7.0$, 1978) in relation to the release of prediction information and aftershock information will be reported in Chapter 11. As demonstrated in this example, the information could be so distorted during its propagation that many people would tend to overreact against the deformed information.

Reaction of the public against an earthquake warning varies from country to country depending upon national traits, economic conditions, and so on. The point that a warning should be issued after knowing possible reaction in detail will be emphasized in Chapter 12. A few examples of such reaction will be covered by the chapter.

Finally, legislations relevant to earthquake prediction and warning will be referred to in Chapter 13. The main topics will be the Large-scale Earthquake Countermeasures Act in Japan and the Earthquake Hazards Reduction Act of 1977 in the U.S.A. Both laws will be cited in the Appendices.

The publication of this book is planned as one of the volumes of the monograph series called "Developments in Earth and Planetary Sciences (DEPS)" published by the Center for Academic Publications

Japan/D. Reidel Publishing Company. Judging from the recent development of earthquake prediction along with the social concern with earthquake warning, the author, who is the General Editor of DEPS, believes that the book is timely. In the course of preparing the manuscripts, the author was assisted by Mr. K. Oshida of the Center for Academic Publications Japan. Miss M. Omi also helped the author through the typing and drawing work. The author is grateful to them. A part of the translating and publishing cost is covered by the grant given to the author by the Ministry of Education, Science, and Culture of Japan for which the author expresses his sincere thanks.

RECENT DEVELOPMENT OF EARTHQUAKE PREDICTION PROGRAMS

A number of countries have been making effort toward earthquake prediction in recenty years. In the People's Republic of China, imminent predictions were made public prior to the 4 disastrous earthquakes having a magnitude larger than 7 that occurred during 1975–76. As a result many lives were saved from otherwise fatal disasters. It is regrettable, however, that no imminent prediction was issued for the 1976 Tangshan earthquake, that struck a highly-populated area in Hebei Province resulting in the loss of 242,000 lives according to the official announcement.

A national program on earthquake prediction has been officially started in the U.S.A. as a part of the Earthquake Hazards Reduction Program (EHRP) in 1973. Intensive observation networks of various disciplines have now been developed in California, Nevada, Aleutian-Alaska, and so on.

Much effort toward earthquake prediction has been concentrated to seismic areas in Middle Asia, Baikal Rift, and Kamchatka in the U.S.S.R. since the 1950's.

The 4th 5-year program on earthquake prediction starting in 1979 was recommended to the Japanese government by the Geodetic Council, responsible for the coordination of geophysical work in Japan, the program being much more extensive than previous programs in various respects. The Large-scale Earthquake Countermeasures Act, which relies largely on long-term as well as short-term earthquake prediction information, was enacted in Japan at the end of 1978. Prediction of a large-scale earthquake, that is supposed to occur sooner or later in the Tokai area, an area between Tokyo and Nagoya in Central Japan facing the Pacific Ocean, has now become a matter of actual necessity.

In view of the development of earthquake prediction programs in respective countries, the author believes that it is desirable to summarize newly-developed programs on earthquake prediction in this chapter. What the author has written comprises rather extensive supplements of the previous summaries (RIKITAKE, 1976a, 1978a) based on modern literature which currently came to his attention.

1.1 Japanese Program

1.1.1 Scope, budget, and development

It was stated in the author's previous text (RIKITAKE, 1976a) that the Japanese program on earthquake prediction had been started in 1965 on the basis of a plan called the "blueprint" of earthquake prediction as proposed by the Research Group on Earthquake Prediction Program in Japan (TSUBOI et al., 1962).

That the program had to be modified later by the experiences of a few strong earthquakes and the social demands for practical earthquake prediction and that the Coordinating Committee for Earthquake Prediction (CCEP) was established have also been stated in the previous text.

In 1976, debate about the possibility of occurrence of a great earthquake in the Tokai area became of social concern for a number of reasons. It was then required by the locale, which is under threat of the quake, to establish a system that handles the imminent prediction of the feared earthquake. Prediction Council for the Tokai area was thus started in April 1977.

In order to meet the requirement for short-term and imminent-earthquake prediction, the program on earthquake prediction, which used to be aimed at long-term prediction only, has had to be modified. Much observed data should now be sent to JMA on the on-line real-time basis. The 4th 5-year program on earthquake prediction as started in 1979 has necessarily become an expensive one. In Table 1.1

TABLE 1.1.   Yearly budget of the Japanese program on earthquake prediction.

Yaear	Amount (million yen)
1965	212
1966	290
1967	334
1968	328
1969	496
1970	596
1971	805
1972	898
1973	761
1974	1,552
1975	2,007
1976	2,312
1977	3,678
1978	4,124
1979	5,847
1980	6,005
1981	6,433
Total	36,678

TABLE 1.2.   Development of the Japanese program on earthquake prediction.

1961
Apr.   The first meeting of the Research Group on Earthquake Prediction Program was held.

1962
Jan.   A pamphlet entitled "Prediction of Earthquakes—Progress to Date and Plans for further Development" (the so-called blueprint of earthquake prediction) was published by the Research Group on Earthquake Prediction Program.

1963
June   Section for Earthquake Prediction was set up in the Geodetic Council, Ministry of Education.

Nov. 7   Science Council of Japan recommended promotion of earthquake prediction research to the Japanese government.

1964
July 10   Geodetic Council proposed to the Japanese Government to put the 1st 5-year research program on earthquake prediction into operation.

1965
Mar.   Sub-committee for Earthquake Prediction was set up in the National Committee for Geodesy and Geophysics, Science Council of Japan.

Apr.   Research program on earthquake prediction was officially started.

Dec. 1   Special Committee for Disaster Prevention of the House of Councillors passed a resolution for intensifying earthquake countermeasures.

4

TABLE 1.2   (continued)

1966	
May	Committee on Crustal Activity related to the Matsushiro Earthquakes was formed.
1968	
May 24	In view of the hazards caused by the Tokachi-Oki earthquake, the Cabinet approved the promotion of earthquake prediction.
July 16	Geodetic Council proposed to the Japanese Government to put the 2nd 5-year program on earthquake prediction into operation.
1969	
Apr.	Coordinating Committee for Earthquake Prediction was set up in the Geographical Survey Institute.
1970	
Feb. 20	South Kanto area is designated by the Coordinating Committee for Earthquake Prediction as the area for intensified observation along with 8 other areas for special observation.
1973	
June 29	Geodetic Council proposed to the Japanese Government to put the 3rd 5-year program on earthquake prediction into operation.
July 6	Central Disaster Prevention Council reached a conclusion that disaster countermeasures should be strengthened.
Aug. 20	Following the agreement at the Central Disaster Prevention Council, a body that coordinates earthquake prediction administration among various ministries involved was formed.
1974	
Feb. 28	Coordinating Committee for Earthquake Prediction designated the Tokai area as an area for intensified observation.
Nov. 7	The body for earthquake prediction administration as set up in Aug., 1973 was changed into a committee headed by the Deputy Minister of the Science and Technology Agency.
1975	
July 25	Geodetic Council proposed to the Japanese Government to revise the 3rd 5-year program on earthquake prediction.
Aug. 15	Central Disaster Prevention Council reached a conclusion for urgent intensification of disaster countermeasures.
1976	
Oct. 29	Headquarters for Earthquake Prediction Promotion chaired by the Minister of Science and Technology Agency was established in the Cabinet.
Dec. 17	Geodetic Council proposed to the Japanese Government to revise again the 3rd 5-year program on earthquake prediction in order to meet the social demands for imminent earthquake prediction based on incessant observations.
1977	
Apr. 4	Headquarters for the Promotion of Earthquake Prediction decided to establish the Prediction Council for the Tokai area in the Coordinating Committee for Earthquake Prediction.
1978	
June 7	Large-scale Earthquake Countermeasures Act passed the Diet.
July 12	Geodetic Council proposed to the Japanese Government to put the 4th 5-year

TABLE 1.2   (continued)

	program on earthquake prediction in operation.
Aug. 21	Coordinating Committee for Earthquake Prediction revised the areas for special observation.
Dec. 14	Large-scale Earthquake Countermeasures Act was put in force.
1979	
June	Prime Minister designated the areas under intensified measures against earthquake disaster basing on the Large-scale Earthquake Countermeasures Act. The areas cover the Shizuoka Prefecture along with portions of neighboring prefectures.

is shown the annual budget defrayed for the earthquake prediction program since 1965. It should be emphasized that the amounts in the table do not include salaries for personnel working on the program.

The development of the Japanese program on earthquake prediction since its very beginning is summarized in Table 1.2 in which some of administrative matters related to the subject are also included.

The main points achieved by the Japanese program on earthquake prediction will be summarized in the following subsections. What will be presented there are confined only to the numbers of observatories increased by the program, the increase in length of levelling routes and number of geodetic stations under the program and the like. Scientific evaluations of such achievements will be presented in later chapters.

1.1.2   Achievements by the program

1)   Geodetic survey

Geodimeters with a laser source are now in general use in horizontal geodetic surveys of current earthquake prediction programs. It has been proposed in the 3rd Japanese program on earthquake prediction that precise geodetic networks are to be set up. The first-order net as composed of the first- and second-order triangulation stations amounting to 6,000 in number is planned to be surveyed every 5 years, and second-order net as composed of 30,000 third-order triangulation stations every 10 years. Such repetitions of geodetic survey are doubtless useful for monitoring accumulation of crustal strain.

6

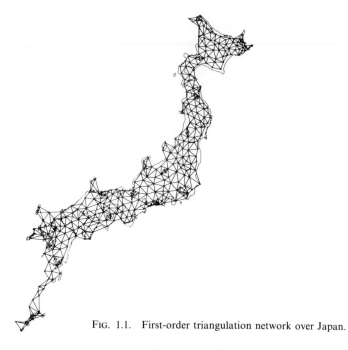

FIG. 1.1.   First-order triangulation network over Japan.

TABLE 1.3.   The numbers of stations for the first-order precise geodetic survey as approved by the Ministry of Finance.

Year	Number of stations
1974	300
1975	415
1976	535
1977	715
1978	985
Total	2,950

In response to the budgetary request of the GSI, the Ministry of Finance approved the expenses only necessary for conducting the surveys for the first- and second-order triangulation stations, the number of which are listed in Table 1.3. According to the original plan, it is clear that 1,200 stations are to be surveyed every year. Although the number of approved stations is approaching that originally planned, only about one half of the proposed survey could be accomplished during the 3rd 5-year program because of the budgetary limitation.

It is therefore stressed that something drastic must be done in the 4th 5-year program in order to increase the budget for the survey of the precise geodetic networks. It is feared, however, that no dramatic increase in the budgetary request can be made because of the balance between GSI's various jobs some of which have nothing to do with earthquake prediction. To make matters worse, the continuing price-rise of commodities makes it difficult to accomplish the survey of all the stations approved.

It has been proposed in the blueprint of the Japanese program on earthquake prediction that the first-order triangulation survey should be repeated throughout Japan every 10 years. In light of the development of geodimeter, it so happened that the program was revised in such a way that trilateration replaces triangulation as stated above.

As for levelling surveys, it was proposed in the blueprint that the first-order levelling route, amounting to 20,000 km in total length, is to be resurveyed with a time-interval of 5 years. The budgetary arrangements for such survey work has been tentatively completed in essence. However, the price-rise of commodities also threatens the

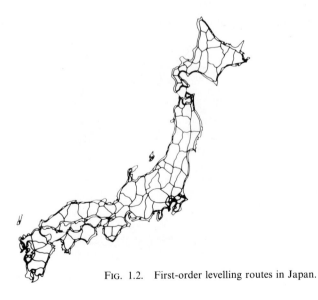

FIG. 1.2.   First-order levelling routes in Japan.

Fig. 1.3. A bench-mark for the second-order levelling survey at Yugashima Town in Izu Peninsula, Shizuoka Prefecture, Japan.

work. The resurvey interval had to be extended. It is strongly stressed in the 4th program, therefore, that the levelling surveys should be conducted as they were planned initially.

2) Tide-gauge observation

About 100 tide-gauge stations were constructed at intervals of approximately 100 km along the coasts of Japanese Islands under the program on earthquake prediction as can be seen in Fig. 1.5. The data from approximately 90 stations are sent to the Coastal Movements Data Center (CMDC) of the GSI and processed there. Hourly values of tide-gauge observations gathered at a number of stations in the Tokai area are telemetered to JMA via GSI on the real-time basis.

Fig. 1.4. Aburatsubo Tide-gauge Station about 60 km south of Tokyo.

FIG. 1.5.   Tide-gauge stations in Japan excluding those on Okinawa and Ogasawara islands.

3)   Gravity

No gravity work was included in the blueprint because it was understood at that time that only very small changes in gravity would be related to an earthquake occurrence. The change, if any, was thought to be so small that it would be hard to detect. It has become clear in recent years, however, that changes in gravity do actually take place in association with a crustal uplift which is sometimes related to an earthquake occurrence. Such a conclusion was reached by the improvement of gravimeters and, at the same time, by the revision of gravity survey techniques.

The gravity changes associated with the 1965–67 Matsushiro earthquakes and the 1970–79 crustal uplift in the Izu Peninsula are very useful for inferring what is going on underneath these areas. The author thinks that gravity measurement will play an important role in understanding the processes occurring at a focal region in the future. Attention is drawn to new transportable instruments capable of measuring the absolute value of gravity with an accuracy of 10 microgal, now in process of being completed.

4) Continuous observation of crustal movement

In order to supplement the crustal movement monitoring by geodetic surveys, which are essentially intermittent, the blueprint proposed to construct 100 observatories for continuous crustal movement equipped with tiltmeters and strainmeters. It turned out, however, that such a plan cannot be put into practice because of the difficulties in finding suitable places for constructing an underground

FIG. 1.6.   Crustal movement observatories equipped with tiltmeters, strainmeters and the like installed in a vault.

FIG. 1.7.   The site where the sensor of a volume strainmeter is buried at Omaezaki weather station facing the Enshunada Sea, off the Pacific coast of Central Japan.

vault. Even if we can find a site, the expense for constructing the vault originally proposed would become prohibitively high.

Figure 1.6 indicates the locations of crustal movement observatories equipped with tiltmeters and strainmeters as proposed in the blueprint, among which 18 observatories were constructed under the program on earthquake prediction.

In view of the difficulties in securing observation sites as stated above, bore-hole tiltmeters and strainmeters of various types have been designed and tested. It appears that the bore-hole volume strainmeters, which were developed by Sacks *et al.* (1971) and adopted extensively by JMA, are working very well. Thirty-one strainmeters of this type have been buried along the Pacific coast in the South Kanto and Tokai areas. The observed data are telemetered to the central office of JMA on the real-time basis. Figure 1.9 shows the locations of these strainmeters.

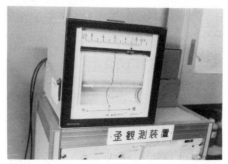

Fig. 1.8.   The recorder of the volume strainmeter in a recording room of Omaezaki weather station.

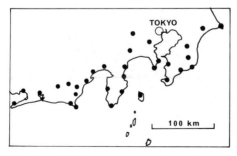

Fig. 1.9.   The distribution of JMA's volume strainmeter stations in the Tokai-Kanto area.

12

5) Seismicity

JMA's routine seismic observation system has been dramatically modernized by the program on earthquake prediction. As can be seen in Fig. 1.10, Japan is covered by a network consisting of some 110 stations, many of which are equipped with visual-recording electromagnetic seismographs and high-sensitivity tape-recording seismographs. It has so far been planned to monitor all shocks having a magnitude of 3 or over by the JMA net. JMA's effort toward the improvement of seismic observation has been so outstanding that the detectability in the Kanto area, an area surrounding Tokyo, actually reached the goal.

In order to avoid noise in association with the industrialization of Japan, use of seismometers buried in a bore-hole has become fashionable in recent years. It has been pointed out, however, that the JMA's seismic net should further be strengthened by making use of telemetering and computer processing of data. These points are certainly under consideration in the 4th 5-year program on earthquake prediction.

Monitoring of microearthquakes ($1 \leq M < 3$) has been made by

FIG. 1.10.   JMA stations for routine seismic observation excluding those on islands south of Kyushu and Ogasawara.

microearthquake observatories mostly belonging to national universities. Nineteen microearthquake observatories having several
satellite seismic stations have been constructed under the program on
earthquake prediction. The number of microseismic stations reached
135 by the end of 1977 as can be seen in Fig. 1.11. Since the 3rd 5-year
program, use of telemetering became popular, so that most seismic
data are nowadays recorded at centers attached to national universities and processed by a computer.

Looking at Fig. 1.11, however, we see that observation nets are
incomplete at the boundaries between each system of observation.
Almost no microseismic observatory has been set up in the Kyushu
area. Use of telemetering is quite backward in the Shikoku area.
These points must be improved in the future program.

Mobile parties for observation of ultra-microearthquakes
($M < 1$) have been attached to a number of national universities.
Whenever an anomalous seismic activity takes place, these parties are
sent to the area with their instruments carried by a motor-car.

FIG. 1.11. Stations for microseismic observation and telemetering systems completed by 1977.

According to the experiences during a number of seismic activities, it has now become recognized that a party of this kind should be equipped with a transportable telemetering system.

6) Sea-bottom seismic observation

No technique that makes a sea-bottom seismic observation possible had been developed when the blueprint was drafted, so that such techniques were omitted in the blueprint about sea-bottom seismographs. It has later become clear, however, that seismicity underneath sea-bottom can be monitored by making use of seismographs of anchored-buoy type and free-dropping and self-floating type.

As social demands for predicting the feared Tokai earthquake became strong around 1977, JMA developed a system of sea-bottom seismic observation that is set up off the Pacific coast of the Tokai area. Signals from 4 seismometers along a cable extended to a distance as far as 110 km off the coast are now monitored at a station at Omaezaki, and they are also telemetered to the JMA's central office in Tokyo on the real-time basis.

7) Deep-well seismic observation

Microearthquake observation in the Tokyo area has become difficult because of the substantial noise caused by industrial activity and traffic. The National Research Center for Disaster Prevention undertook deep-well observations hoping to avoid the noise by putting seismometers in the basement rock underneath the Kanto plain covered by a thick sediment. The first observation point was set

FIG. 1.12. Deep-well seismic station of the National Research Center for Disaster Prevention at Iwatsuki City, Saitama Prefecture.

up at Iwatsuki City, Saitama Prefecture, about 28 km north of Tokyo where seismographs were placed at a depth of 3,510 m. It then turned out that the noise is so greatly reduced at this depth that an observation, which is more sensitive than that at the ground surface by a factor of 1,000, can be achieved.

The second well for the same purpose was dug at Shonan Town, Chiba Prefecture, a few tens of kilometers to the east of Tokyo. The depth is 2,300 m in this case. In 1980, the third well of 2,700 m in depth was completed at Fuchu City at the western part of Tokyo about 40 km distant from its center.

Combining this observation net with ERI's net, the accuracy and detectability of microearthquake observation in the Tokyo area increased tremendously. Should foreshocks, that were said to be observed prior to the Ansei Edo earthquake ($M = 6.9$, 1855), which occurred immediately beneath Tokyo, occur some time in the future, they are capable of detection by the system of deep-well seismic observation.

8)   Changes in seismic wave velocities

Since the beginning of the program on earthquake prediction, it has been attempted to detect changes in seismic weve velocities by making an explosion on Izu-Oshima Island, about 100 km south of Tokyo, and observing arrivals of seismic waves at a number of stations on the main island of Japan.

Explosions of about 500 kg explosives have been repeated more than 10 times approximately with an interval of 1 year. A few other shot-points were added to that on Izu-Oshima Island in these years. No significant changes in seismic wave velocities have been found by these observations.

9)   Geomagnetism and earth-currents

Digital-recording proton precession magnetometers have been set up at the stations shown in Fig. 1.13. Magnetic surveys have also been intensified. The pattern of geomagnetic secular variation in Japan has now been established very clearly. There are areas such as Izu Peninsula where we observe anomalous secular variations which differ considerably from the general tendency.

A resistivity variometer of high sensitivity has been developed by YAMAZAKI (1975). The variometer, that has been in operation at the

FIG. 1.13. Stations equipped with a proton precession magnetometer.

Aburatsubo Crustal Movement Observatory about 60 km south of Tokyo, often records coseismic as well as preseismic resistivity changes.

An attempt to detect changes in ground resistivity, if any, has been tried in recent years by making an electric current flow into the ground from a pair of electrodes, the distance between which amounts to the order of 1 km. In spite of Chinese and Soviet reports, no outstanding success in monitoring changes in resistivity prior to an earthquake occurrence has been achieved. Accurate observations of this kind are difficult to make in Japan because of electrical stray currents from railways and factories operated by d.c. This is also the case for observation of changes in natural potential difference between two electrodes buried in the ground. Nevertheless a few observations of anomalous changes in earth potential probably related to seismic activity have recently been reported.

The ratio of amplitude of the vertical component to that of the horizontal component usually takes on an approximately constant value in the case of short-period geomagnetic variations observed at a middle-latitude station. Such a ratio, that is called the transfer function, seems to be subjected to a change prior to an earthquake in some cases. A number of examples which indicate preseismic changes in the transfer function were recently presented.

It has sometimes been reported that the amplitude of short-period geomagnetic variations is subjected to a change before an earthquake. Such a change can only be confirmed by comparing the magnetograms at a certain station to that at a standard station which is fairly distant from the earthquake area.

Many of the above items related to geomagnetism and geoelectricity were not included in the blueprint. Although each item is still in a preliminary stage of research, we may well expect future developments of geomagnetic and geoelectric approach to earthquake prediction.

10)   Active fault

A program aiming at examining active faults all over Japan has been energetically carried out especially under the 3rd program on earthquake prediction. As a result the distribution of active faults is brought to light as shown in Fig. 1.14 (RESEARCH GROUP FOR ACTIVE FAULTS, 1980), and, at the same time, the rating of activity of each fault is estimated.

Geological Survey of Japan published maps of active facult with various scales. The Hydrographic Office of the Maritime Safety Agency produced submarine maps around the Japanese Islands, many faults being indicated on the map.

Geophysical and geological investigations of various disciplines were concentrated on the Yamasaki fault in the Hyogo Prefecture. There is reason to believe that the fault is active, so that an earthquake of moderate magnitude may well be expected to occur in the near future. Exercises of earthquake prediction at such a test field would certainly be a useful preliminary towards actual prediction.

11)   Rock breaking test

Since the very start of the program on earthquake prediction, it has been deemed that experimental studies on rock failure and associated phenomena are important because they would provide the background of earthquake prediction theory. High-pressure apparata that can be used for rock breaking tests of various kinds were installed at national universities and the Geological Survey of Japan under the 3rd 5-year program on earthquake prediction.

18

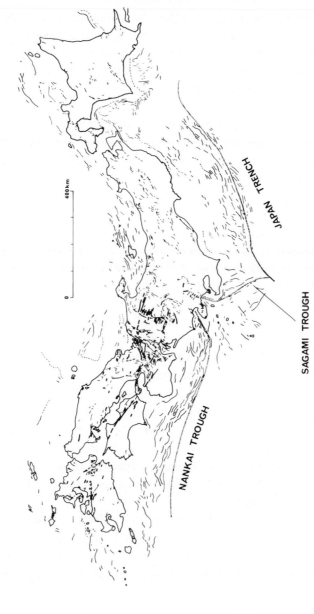

FIG. 1.14.   Active faults in and around Japan. Thin and thick lines indicate dip-slip faults and active faults having lateral strike slip component. Broken lines show monoclines beneath the sea (RESEARCH GROUP FOR ACTIVE FAULTS, 1980).

12) Underground water

Observation of underground water has not been included in the blueprint because it was believed at that time that the relation between underground water and earthquake is so complicated that no effective analysis can be made.

In view of many Soviet and Chinese reports on precursory changes in the level and quality of underground water, observation of underground water has recently become an official subject of the program on earthquake prediction.

In association with an anomalous uplift in Kawasaki City in 1974, the Geological Survey of Japan dug a well having a depth amounting to 1,016 m and began a continuous observation of radon content. The same institute further dug 6 wells in the Tokai and Izu areas in 1976–77. In addition to these wells, many existing wells are used for monitoring underground water. Observations at a regular interval are made at 40 wells, while various elements of undergound water at 10 wells are being telemetered to JMA via the Geological Survey of Japan.

A number of national universities also began underground water observations in the Tokai and Izu areas towards the end of the 3rd program on earthquake prediction. Special funds were defrayed from the Science and Technology Agency to an amateur group in 1978 in order to encourage underound water monitoring by the group.

13) Historical document

Characteristics of earthquake occurrence in a certain area may be inferred to some extent by analysing historical documents on large earthquakes, should they be available. Since the start of the 3rd 5-year program on earthquake prediction efforts toward collecting historical documents related to earthquakes has been made with special funds.

14) Anomalous animal behavior

It has long been said in Japan as well as foreign countries that animals behave anomalously before an earthquake. Towards the end of the 3rd 5-year program, special funds were raised for tackling the problem by the Ministry of Education. Y. Suyehiro, a professor emeritus at the University of Tokyo and a famous ichthyologist, formed a research group with the funds. Although it is still not known why animals behave anomalously and what stimulation makes them excited, statistics of precursor time, the interval between the time when an animal behaves anomalously and that of the main shock, indicate that animal precursors, should they be earthquake precursors at all, have precursor times centering at about 0.5 day and 2–3 hours (RIKITAKE, 1978b, 1979a, 1981b; RIKITAKE and SUZUKI, M., 1979).

1.2 U.S. Program

1.2.1 Scope, budget, and development

The author, who has organized the U.S.-Japan seminars on earthquake prediction problems in the past, spent considerable time in the U.S.A., and became acquainted with many American scientists at various meetings, believes that he is familiar with the U.S. program on earthquake prediction to some extent. But the recent advance in the program is so fast that it is extremely difficult to follow all the most up-to-date developments of the program. Although the author will attempt in the following to summarize the earthquake prediction program in the U.S.A., he hopes that something more advanced will be added to what he writes. He will be happy if American colleagues would correct or supplement that presented in the following. The author refers extensively to PRESS et al. (1965), WALLACE (1974), RIKITAKE (1974), HAMADA (1975), PANEL ON EARTHQUAKE PREDICTION OF THE COMMITTEE ON SEISMOLOGY (1976), WARD (1979), WYSS (1981), and MacCABE (1979).

As was stated in the author's previous text (RIKITAKE, 1976a) prediction-oriented seismology was started in the United States after the 1st U.S.-Japan seminar on earthquake prediction in 1964. The disastrous Alaska earthquake ($M=8.4$, 1964) that occurred soon after the seminar doubtless gave rise to public concern about earthquake prediction.

PRESS et al. (1965) proposed a 10-year program on earthquake prediction, of which the cost for the work in the California and Nevada areas was estimated as 30,400,000 U.S. dollars. Although the program has never been implemented probably because of financial difficulties, it is certain that the program stimulated interest in earthquake prediction among the seismological community. Since around 1965, the number of seismologists and geophysicists who carry out earthquake prediction research began to increase considerably in the U.S.A. Most of their researches were supported by the National Science Foundation (NSF) and the United States Geological Survey (USGS).

Reorganization of federal institutions dealing with solid earth sciences took place in 1973. On that occasion, a program on earthquake prediction and earthquake engineering was decided to start as one of the major efforts of Federal Government. Being called the Earthquake Hazards Reduction Program (EHRP), the program involved USGS, NSF, National Oceanic and Atmospheric Administration (NOAA), Atomic Energy Commission (AEC), and the Department of Housing and Urban Development (HUD).

Much earthquake prediction work is handled by USGS, while NSF is responsible for most work related to earthquake engineering.

The author understands that the EHRP budget for the 1974 fiscal year (FY) amounted to some $ 19,400,000. This amount indicates that the budget for earthquake problem almost doubled as compared to that for FY 1973. The amount of EHRP money available for earthquake prediction was of the order of 5 million dollars.

The budget increased discontinuously in 1978 because of the enactment of the Earthquake Hazard Reduction Act of 1977 and

TABLE 1.4. Earthquake Hazards Reduction Program FY 1978 funding as enacted (MacCabe, 1979).

Program element	USGS	NSF	Percentage of program
	(thousands of dollars)		
Fundamental studies	2,650	5,300	14.9
Prediction	15,764*	—	29.4
Induced seismicity	1,200	—	2.2
Hazards assessment	10,607	—	19.8
Engineering	—	15,500	29.0
Research for utilization	—	2,500	4.7
Totals	30,221	23,000	

*   Includes $1,500,000 specifically for prediction studies in foreign countries.

subsequent legislation. As can be seen in Table 1.4, the budget of the whole EHRP enjoyed a three-fold expansion, and so the budget for prediction reached an amount of the order of 15 million dollars.

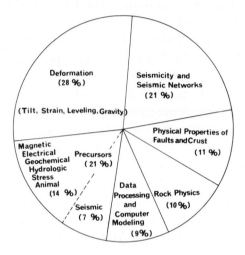

FIG. 1.15.   Allocation of funds to different types of prediction studies during FY 1978 (MacCabe, 1979).

TABLE 1.5.   Allotment of USGS funds FY 1977 for prediction (internal and contract program) to different disciplines.

Discipline	Budget in $1,000	Percentage
Passive seismology	2,730	37
Tilt measurements	965	13
Geodetic surveys	959	13
Rock mechanics	780	11
Geochemical techniques	401	5
Theoretical modeling	386	5
Stress measurements	259	4
Magnetic surveys	244	3
Electrical methods	193	3
Creep measurements	189	3
Strain measurements	147	2
Animal studies	36	0.5
Socioeconomic studies	25	0.5
Total	7,314	100.0

Allocation of funds to different types of prediction researches during FY 1978 is shown in Fig. 1.15. A more detailed allocation of USGS funds to various disciplines for FY 1977 was presented by P. Ward (unpublished, 1977) at the 1977 U.S.-Japan seminar on earthquake prediction as shown in Table 1.5. It is interesting to note that both the animal and socioeconomic studies are officially mentioned although the funds allotted to them are not large. It is very remarkable that a fairly large portion of the budget, as much as 37%, is allocated to seismic observation.

In Table 1.6 (Ward, unpublished, 1977) is shown the budget classified according to the nature of expenditure. As the salaries and

TABLE 1.6.   The breakdown of USGS funds FY 1977.

Nature of expenditure	Budget in $1,000	Percentage
Salaries	2,852	39
Overhead	2,121	29
Other expenses	2,341	32
Total	7,314	100

overhead occupied a large proportion, the amount that can be used for actual observations and researches is substantially smaller than that of the Japanese program even though the discontinuous increase in budget in 1978 is taken into account.

About one-third of the USGS budget for earthquake prediction is distributed among researchers outside the USGS, namely a sum of about $ 5 million is given to researchers in universities, state agencies, and private sectors for FY 1978. It has been criticized, however, that the funds should be handled by NSF rather than USGS which spends about two-third of the budget of earthquake prediction. Although a committee consisting of eminent scientists outside USGS is responsible for allocation of funds, some people feel that it is not fair that USGS controls the budget.

1.2.2   Achievements by the program

According to MacCabe (1979), the U.S. earthquake prediction program consists of over 100 projects funded in a variety of uni-

versisties and industries and within the USGS. It is extremely difficult
to summarize the achievements by such a diverse program, so that
only a few main points that come to the author's attention are
mentioned below.

1) Seismic observation

As can be seen in Table 1.5, a major portion of prediction
program budget has been spent on constructing microearthquake
observation arrays mostly over California. According to J.P. Eaton
(KISSLINGER and RIKITAKE, 1974), since 1973, an array consisting of
102 seismometers has been developed by USGS over a 350×50 km
area covering the northern portion of the San Andreas fault in
Central California. The observed data telemetered to the National

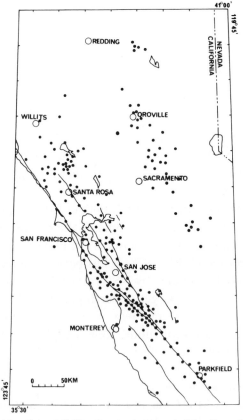

FIG. 1.16.   Central California seismic network (MACCABE, 1979).

Center for Earthquake Research (NCER) in Menlo Park. In 1978, some 200 seismic stations are in operation as can be seen in Fig. 1.16.

The scope of observation has since been extended. In South California, an array of 150 short-period seismometers has by 1976 been completed by the California Institute of Technology (CIT) in cooperation with USGS as can be seen in Fig. 1.17. Observed data are

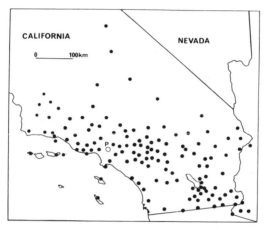

FIG. 1.17.   South California seismic network. P: Pasadena (WHITCOMB, 1977).

telemetered to CIT and processed by a processing system called the CEDAR (Caltech Earthquake Detection and Recording) (WHITCOMB, 1977).

The seismic observation has now been extended to northern California (J. P. Eaton, personal communication, 1979).

2)   Geodetic work

About 1,200 geodimeter lines crisscrossing the faults have been set up in California and Nevada. The average length of the lines amounts to 20 km, while the accuracy of length measurement is estimated as 5×10^{-7}. About 500 lines are resurveyed every year.

First-order levelling routes about several thousand kilometers in the total length have been established in California and Nevada. It is of concern, however, that the regular resurvey intervals are longer than 5 years, so that levelling surveys cannot be adequately applied to earthquake prediction. In spite of such a defect, it is somehow found by CASTLE et al. (1976) that an anomalous uplift amounting to 25 cm at maximum took place during 1960–75. The uplift centering at Palmdale, about 50 km north of Los Angeles, covered a 400×150 km area. As the area includes the portion of the San Andreas fault which ruptured at the time of the 1857 Fort Tejon earthquake, of which the magnitude was estimated to reach a value around 8, it is feared that the uplift might have something to do with an occurrence of a great earthquake. Special funds amounting to 600 milliom dollars were raised in 1976 for the purpose of clarifying the nature of the uplift. Various observations have since been made over the area in question with these funds, so that the area became a target of earthquake prediction operation.

It turned out, however, that a bench mark about 10 km south of Palmdale subsided by 18 cm during 1964–76 as shown in Fig. 1.18 (SAVAGE and PRESCOTT, 1979). Meanwhile the uplifted area seems to

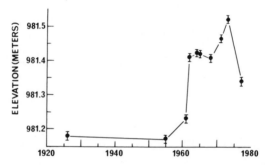

FIG. 1.18.   Changes in elevation of a bench-mark, 10 km south of Palmdale, relative to the tide-gauge station in Los Angeles (SAVAGE and PRESCOTT, 1979).

extend to the California-Arizona border as can be seen in Fig. 1.19 (MACCABE, 1979). It is not at all clear whether the anomaly may be linked to an impending earthquake. SAVAGE and PRESCOTT (1979) showed that no anomalous strain accumulation is taking place over the uplifted area. CASTLE (1978) argued that there is some evidence that a similar period of uplift and partial collapse may have taken place between 1902 and 1926. It appears to the author that there are a number of American seismologists and geophysicists who are now inclined to doubt that the uplift is connected with an occurrence of a major earthquake. Much debate about the reality of the uplift was presented at the Southern California session of the fall 1980 John

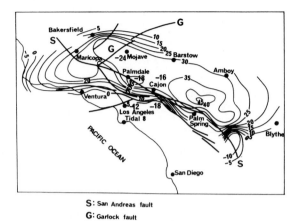

S: San Andreas fault

G: Garlock fault

FIG. 1.19.   Area of recent uplift in the Transverse Range and adjacent parts of the Mojave Desert in California during the 1959–74 period. Uplift occurs along the San Andreas fault. Changes in elevation are shown with contours of 5 cm interval. Larger numerals indicate subsidence in cm since early 1974 (MACCABE, 1979).

Muir Geophysical Society (RUNDLE and MCNUTT, 1981). It seems likely that the error of levelling due to atmospheric refraction may sometimes be fairly large. It was argued that most of the measured uplift went away when refraction corrections were applied.

3)   Tiltmeter, strainmeter, and creepmeter observation

USGS pays much attention to tiltmeter observation for possible earthquake precursors (MORTENSEN and JOHNSTON, 1975). The tiltmeter sites along the San Andreas fault system are shown in Figs. 1.20 and 1.21 respectively for the northern and southern parts of California (MACCABE, 1979).

According to WYSS (1981), USGS set up about 60 tiltmeter stations in California and about 20 in Alaska. USGS also gives university workers about 40 tiltmeters although university people did not so request. Many geophysicists are not in favor of the use of a USGS-type tiltmeter, which is buried in the ground at a depth of only several meters and works on the principle of bubble motion because it is likely to be affected very seriously by meteorological conditions. Nevertheless precursor-like changes in tilting have sometimes been reported by USGS (e.g. JOHNSTON, 1978b).

USGS has been operating 11 strainmeters on the San Andreas

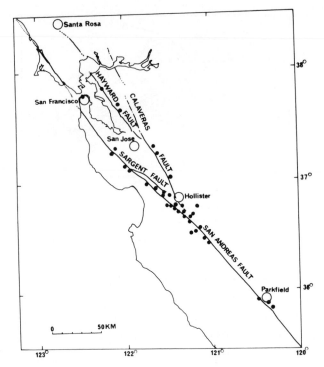

FIG. 1.20.   Tiltmeters sites in northern California (MACCABE, 1979).

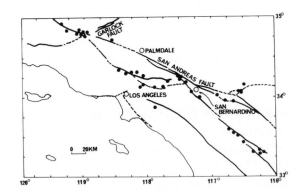

FIG. 1.21.   Tiltmeter sites in southern California (MACCABE, 1979).

fault since 1974. No signal that is thought to have something to do
with a possible earthquake precursor has been found (JOHNSTON,
1978c).

In addition to tiltmeters and strainmeters, more than 10 creepmeters are set up along the San Andreas fault mainly in Central California. They are constantly monitoring the fault movement (BURFORD *et al.*, 1978).

4)   Gravity

A few hundred gravity stations are set up over the uplifted area of southern California as are shown in Fig. 1.22 (MacCABE, 1979).

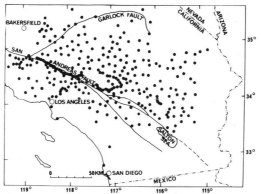

FIG. 1.22.   Location of gravity stations in southern California (MacCABE, 1979).

Effort toward producing a portable absolute gravimeter with microgal sensitivity has also been made.

5)   Geomagnetism

Fifteen proton precession magnetometers have been set up along the San Andreas fault in northern California as can be seen in Fig. 1.23 (MacCABE, 1979). Total geomagnetic intensity values measured with these magnetometers are telemetered to NCER at Menlo Park every minute (JOHNSTON *et al.*, 1976; SMITH and JOHNSTON, 1976). The number of magnetometers increased to 21 (JOHNSTON, 1978a) in 1978.

A similar array of proton precession magnetometers will be set up in southern California as can be seen in Fig. 1.24 (MacCABE, 1979).

More than 100 magnetic stations have been provided along faults in California and Nevada. A differential method, which measures the difference in the total geomagnetic intensity between two neighboring stations by making use of a synchronizing technique by radio signal, has been developed (JOHNSTON *et al.*, 1976). A similar

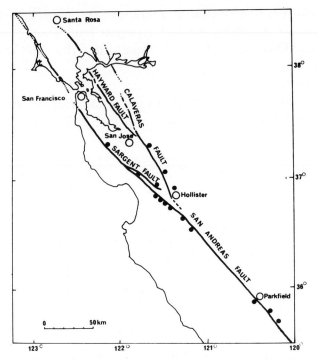

FIG. 1.23. Magnetometers sites in northern California (MacCabe, 1979).

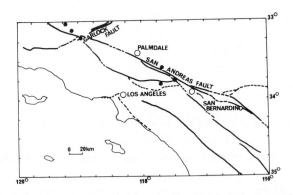

FIG. 1.24. Magnetometer sites in southern California (MacCabe, 1979).

array of magnetic stations is going to be set up over southern California as shown in Fig. 1.25 (MacCabe, 1979).

6) Geoelectricity

Geoelectric observations for resistivity change and earth poten-

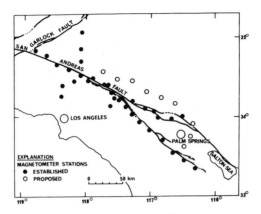

FIG. 1.25.   Magnetometer stations in southern California (MacCabe, 1979).

tial have been carried out by USGS (Bufe *et al.*, 1973), MIT (Massachusetts Institute of Technology) (Fitterman and Madden, 1977), University of California, Berkeley (Mazzella and Morrison, 1974; Corwin, and Morrison, 1977; Morrison *et al.*, 1979), and CIT (Reddy *et al.*, 1976) on the San Andreas fault.

7)   Geochemical study

Monitoring of radon, helium and similar gases at wells and hot springs in relation to seismic activity has been conducted at a number of organizations. For instance, King (1978) reported on a series of weekly measurements at 30 sites on the San Andreas fault. Similar studies have also been made in the Blue Mountain Lake area, New York. Figure 1.26 (MacCabe, 1979) indicates the thermal springs and

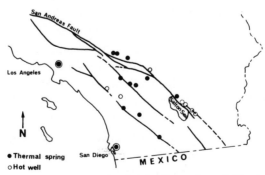

FIG. 1.26.   Radon and helium monitoring sites in southern California (MacCabe, 1979).

hot wells where radon and helium monitoring is carried out by the University of California, San Diego.

8) Underground water monitoring

Not many studies have been made on the relationship between changes in underground water-level and earthquake occurrence in the United States. KOVACH et al. (1975) examined the water level at a well having a depth of 150 m on the San Andreas fault. A few institutes are working on observation of well water-level in southern California (MacCABE, 1979).

9) Other studies

In addition to the prediction-oriented studies above, laboratory experiments on rock samples, in situ measurement of crustal stress and observation of anomalous behavior of animals have been conducted as parts of the earthquake prediction program.

1.3   U.S.S.R. Program

The author meets difficulties in summarizing the Soviet program on earthquake prediction because of a lack of information in recent years. This is probably caused by the linguistic difficulty. No outstanding addition to what the author wrote previously (RIKITAKE, 1976a) can therefore be possible in the following.

It seems likely that an intensive program on earthquake prediction is underway in the Republics of Kazakh, Kirghiz, Uzbek, Tajik and Turkmen, Baikal Rift area, and Kamchatka. It is well known that earthquake prediction work of various disciplines have been conducted in the Garm area, the Republic of Tajik. The work covers disciplines such as geodetic survey, seismic activity, change in seismic wave velocities, source mechanism, ground resistivity, geomagnetic field, and so on. It is well known but worth recalling that the first observation of a drop of the ratio of P wave velocity to S wave velocity foreruns an earthquake (SEMYENOV, 1969) was made in the Garm area. Although much doubt about the general validity of such changes in seismic wave velocities has currently been raised through observation in the U.S.A., Japan, and elsewhere, Semyenov's work is important at least historically.

The report that the ground resistivity in a focal area becomes small by 10–20% prior to an earthquake (BARSUKOV, 1972) stimulated similar work in the U.S.A., Japan, and China. In the Soviet Union, it has been planned to make an electric current as strong as 1,000 amperes flow into the ground by making use of MHD generation. In that case the currents may possibly reach a depth of several tens of kilometers, so that changes in resistivity at the focal region can possibly be monitored.

That the radon content of the well water changed very markedly forerunning the Tashkent earthquake ($M=5.5$, April 26, 1966) was cited in the author's previous text. It is understood that radon measurement has recently been carried out at more than 30 wells in the Tashkent and Fergana areas. Similar observations are also carried out at a number of sites on the Surkob fault system.

The author gathered from Soviet colleagues that about 20,000,000 rubles are spent annually for the earthquake prediction program. About 1,000 scientists, observers, and technicians are devoting more than 50% of their working time to the program. It is also said that two earthquakes of magnitude 7 or so were successfully forecast in 1978. They occurred in Pamir and the Soviet-Iran border, respectively.

1.4   Chinese Program

That an extensive program on earthquake prediction is underway in the People's Republic of China has been brought to light in recent years by the reports of Japanese, American, Canadian, and other seismological missions which visited China. A few Chinese missions that visited Japan, the U.S.A., and so on also talked about the Chinese program fairly freely especially after the purge of the gang of four. Important literature for looking into the Chinese program will be the following: COE (1971), WILSON(1972), BOLT (1974), LEE (1974), OIKE et al. (1975), ALLEN et al. (1975), RIKITAKE (1975a), WHITHAM et al. (1975), ADAMS (1975), RALEIGH et al. (1977), SEISMOLOGICAL SOCIETY OF JAPAN (1976, 1978), SHIZUOKA PREFECTURE (1978), OIKE (1978a, 1979), SUZUKI (1978a, b, c, d), and RIKITAKE et al. (1979).

In addition to the above literature, the author was able to learn much about the Chinese effort toward earthquake prediction through personal talks with members of Chinese seismological delegations which visited Japan and the U.S.A. and with Chinese colleagues on the occasion of UNESCO meetings and symposium. The visits to China in 1978, 1980, and 1981 were especially useful for the author to learn the latest techniques used in China. As the Chief Delegate of the 1978 and 1980 Japanese Seismological Mission and a member of the organizing committee for the proposed 1982 International Symposium on Continental Seismicity and Earthquake Prediction to be held in China, the author could visit Peking (Beijing) and its vicinity, Tangshan, Shanghai, Liaoning, and Sichuan Provinces. In spite of these contacts with Chinese programs, it cannot be said that the author completely understands the earthquake prediction program in China. Nevertheless the author tries to summarize the Chinese program in the following because he believes that a Japanese, able to read Chinese characters to some extent, can possibly offer useful information to occidental colleagues.

According to COE (1971), the Chinese program on earthquake prediction was started soon after the magnitude 6.8 and 7.2 earthquakes that occurred at Xingtai about 300 km southwest of Peking (Beijing) in March 1966. Probably such promotion of earthquake prediction work was put forward on the initiative of the late Premier Chou En-lai who had inspected the earthquake area immediately after the shocks.

ALLEN et al. (1975) stated that several hundred scientists and several thousand technicians are working at about 250 seismic observatories and about 5,000 observation points in China aiming at actual earthquake prediction. Almost all geophysical work except geophysical exploration for underground resources is concentrated on earthquake prediction.

It is clear that the level of Chinese geophysics was not high with few researchers and that the work was theoretical rather than experimental around 1960. It is hardly understood by foreigners why the Chinese undertook such an extensive program starting from an extremely poor stage of earthquake research. Yet, the Chinese have succeeded in issuing imminent earthquake prediction in a number of cases of destructive earthquake.

It is especially of importance and interest to see the Chinese system for earthquake prediction and also to know how administrators handle earthquake prediction information. These points will be dealt with respectively in Chapters 6 and 10. In this section only the Chinese program on earthquake prediction will be reviewed below.

Much of earthquake prediction research in China is made at the Geophysical Institute and the Geological Institute. These two institutes used to belong to the Academia Sinica. But in 1978 main sections of these institutes related to earthquake prediction were transferred to the State Seismological Bureau, and only some divisions for basic research have remained in the Academia Sinica.

The Geophysical Institute was established in 1949. In 1960 and 1966 respectively, the divisions of geophysical exploration and meteorology left the institute because of the new establishment of respective organizations. At the moment, therefore, the Geophysical Institute covers seismology and geomagnetism.

About 400 members are working in the institute; 20 senior and junior researchers (equivalent to professors and associate professors in western countries), 110 research associates, 100 assistants, and other members in the administrative section. There are 8 research divisions in the institute:

Divisions 1 and 4: Seismicity around Peking (Beijing)

Division 2: National seismicity

Divisions 3 and 8: Source mechanism and internal structure of the earth

Division 5: Geomagnetism and earth currents

FIG. 1.27.   Construction of bore-hole seismometer at the Geophysical Institute, State Seismological Bureau in Peking (Beijing).

Division 6: Manufacturing of instruments for observation

Division 7: Scientific information

According to ALLEN (1975), the following are considered the important items of the Chinese program on earthquake prediction:

a) Seismicity patterns—migrations and temporal variations in activity

b) Seismic wave velocities—variations in P and S

c) Radon in well water

d) Telluric currents

e) Geomagnetic field

f) Earth tilt

g) Earth strain

h) Vertical land deformation—levelling surveys

i) Horizontal land deformation—triangulation and trilateration surveys

j) Ground temperature

k) Water level in wells

l) Bubbling, mudding, and swirling of water

m) Animal behavior

Almost all the above disciplines can be covered by the Geophysical and Geological Institutes. The division at the latter institute that is closely related to earthquake prediction is the division called the "seismogeology" where the relationship between the movement of tectonic structural lines and active faults and the earthquake occurrence is studied.

In order to monitor seismicity around Peking (Beijing), 21 seismic

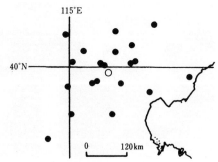

FIG. 1.28.   Seismometer sites around Peking (Beijing) which is shown by an open circle. Signals are telemetered to Peking (Beijing).

stations are in operation as can be seen in Fig. 1.28. Seismograms from these stations are telemetered to the observation center in the building of the Academia Sinica, Peking (Beijing). The author, who visited Peking (Beijing) in 1978, was told that the center of this telemetering system will move to the new site of the Geophysical Institute in the near future. Judging from the fact that they had only 8 seismic stations in 1974, it can be said that the seismic observation in the Peking (Beijing) area has been strengthened at an immense speed.

A part of the telemetering network is also used for telemetry of geomagnetic data. As can be see in Figs. 1.29 and 1.30, signals from 7 proton precession magnetometers are being telemetered to the central observation room in the building of the Academia Sinica as of September, 1978. The author was told that 9 more stations will shortly be added to the existing system. All the instruments for the

FIG. 1.29. Receiving station of telemetered signals from 7 proton precession magnetometers in the Academia Sinica, Peking (Beijing).

FIG. 1.30. Solid circles indicate proton precession magnetometer sites around Peking (Beijing). Signals are telemetered to Peking (Beijing) from these stations.

geomagnetic observation are produced in China. In addition to the above geomagnetic observation system, geomagnetic surveys of total intensity are repeated at 116 magnetic stations over the Peking (Beijing) area at a time-interval of one month.

Earthquake prediction observations in provincial areas are carried out by provincial seismological bureaus, seismological brigades, and seismological teams according to the extent of observation. There are classes of observatories. It is said that there are 17 standard seismological observatories which are responsible for seismic, geomagnetic and geoelectric, crustal movement, gravity, and geochemical observations. At these standard observatories quite extensive observations are usually carried out. In addition to these, there are more than 300 seismic observatories throughout China. Observations at these observatories are not as extensive as those at standard observatories although instruments necessary for fundamental observations are provided. There are also numerous observation points which are equipped with much limited apparata. In association with occurrences of a large earthquake, it is customary to set up temporary field stations.

In addition to the above professional observatories and stations, countless observation stations are operated by amateurs. Some of the amateur stations are set up in electric generation factories, gear-producing factories, brigades for geophysical prospecting, schools, and the like. The workers at these organizations usually have high standard of technology, so that some amateur observatories are as good as the professional. During the 1980 visit, however, the author was told that the number of amateur stations is decreasing. It appears that only stations of high quality survive and that those stations are paid by the government for their work.

In order to see the extent of Chinese seismic observation in a rural area, the distribution of seismic stations in Sichuan and Yunnan Provinces are illustrated in Fig. 1.31 (OIKE, 1978a, b, c).

Figure 1.32 shows the Peking (Beijing) Standard Seismological Observatory at Paichiatang in the western suburb of Peking (Beijing). There are 34 members in the observatory, 26 being scientists and technicians. The observatory is equipped with seismographs of long, medium, and short periods, three-component strong motion seismo-

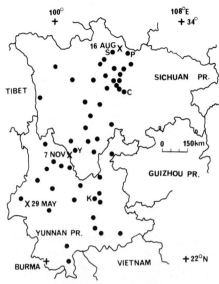

FIG. 1.31. Observation stations in Sichuan and Yunnan Provinces for earthquake prediction. Epicenters of large earthquakes listed in Table 1.7 are also shown with crosses (OIKE, 1978c). C: Chengdu, K: Kunming, P: Pingwu, S: Songpan, Y: Yanyuan.

FIG. 1.32. Peking (Beijing) Standard Seismological Observatory and Chou Jin-ping, the Director.

graphs, proton precession magnetometers, a magnetic theodolite of Schmidt type, an earth-inductor, QHM, optical variometers made in China and East Germany, and so on. The facilities for geomagnetic observation are just comparable to those of the first class observatory in other countries. Measurements of radon content in well water and earth-currents are also routinely carried out there.

40

FIG. 1.33. Room for absolute geomagnetic observation of the Peking (Beijing) Standard Seismological Observatory.

FIG. 1.34. Calibration of field proton precession magnetometers at the Peking (Beijing) Standard Seismological Observatory.

FIG. 1.35. Chengdu, Sichuan Province, Standard Seismological Observatory.

Figure 1.35 shows the Chengdu, Sichuan Province, Standard Seismological Observatory where observations by seismographs of long and short periods, measurement of earth potential and earth resistivity, gravity, earth tilt and three geomagnetic components, and so on are being carried out.

The Pingwu Observatory, Sichuan Province, as shown in Fig. 1.36 is a typical local observatory equipped with a standard seismograph, a radon meter, an earth-current meter, and the like. The observatory was established in 1975. At the time of the 1976 Songpan-Pingwu earthquakes ($M = 7.2$, 6.7, 7.2), the observatory played an important role in promoting earthquake prediction along with the temporary observatories at Shuijing, Gucheng, Wenchuan, and others.

The amateur stations, which the author visited, were generally well equipped. Figure 1.37 is the variometer for geomagnetic declination at a station at the Min Jiang gear factory about 95 km north of Chengdu. The deflection of a small magnet is detected by an optical lever method. The light reflected by a mirror attached to the magnet is received by a photo-transistor. The cover of the magnetometer is

FIG. 1.36.   Pingwu Observatory, Sichuan Province.

FIG. 1.37.   Variometer for geomagnetic declination at an amateur station at the Min Jiang gear factory, about 95 km north of Chengdu.

made of porcelain. In addition to the magnetic observation, observations by short-period seismographs, earth-current meters, carbon-microphone strainmeters, earth-sound meters, and tiltmeters are being carried out by the factory station.

At an electric generation factory at Dujiang Yan, about 50 km northwest of Chengdu, a pair of chemical balances are used for measuring earth tilt as can be seen in Fig. 1.38. It is highly doubtful

FIG. 1.38.   Balance tiltmeters at an electric generation factory at Dujiang Yan, about 50 km northwest of Chengdu.

that such a simple instrument set up in a building, that is surely subjected to a considerable temperature change, can be applied for detecting a precursory tilting. Observations of many other disciplines, such as microearthquake, geomagnetic declination, earth-currents, radon, earth-sound, and the like are also made there.

One of the most outstanding features of the Chinese earthquake prediction observation is the fact that "macroscopic effects" as called by the Chinese are emphasized. The macroscopic effects mean the phenomena that can be observed by the eye and ear without using any sophisticated instruments. Typical examples of such effects are as follows; lowering of water level at a spring, fire-balls that appear from the ground, bubbling at a pond, and anomalous animal behavior. At the time of the Haicheng, Liaoning province, earthquake ($M = 7.3$, 19 h 36 m February 4, 1975), water came out from a hand-pumping well at a commune in the focal area at about 8 o'clock in the morning. Around noon on that day, water began to splash at a height of 1 m. When the author visited there, villagers got together around the well

Fig. 1.39.   A well at a commune in the focal area of the 1975 Haicheng earthquake. Water splashed up from the well 7–8 hours prior to the main shock.

and told us that such an extraordinary thing actually happened as can be seen in Fig. 1.39.

The nation-wide earthquake prediction program in China leads the Chinese to a remarkable success in actual prediction. As can be seen in Table 1.7, they succeeded in issuing an imminent prediction for destructive earthquake four times out of five shocks that occurred during 1975–76. We must congratulate our Chinese colleagues although it is regrettable that no imminent prediction could be made in the case of the Tangshan, Hebei Province, earthquake ($M = 7.8$, 1976) resulting in a large number of deaths. It is not clear why the Chinese failed to issue an imminent warning in this case because long- and medium-term premonitory effects have been exampled and advised very clearly. Unlike the other successful cases, the Tangshan area is highly industrial and so man-made noise must have been high. It is also said that the subversive activity of the gang of four could be one of the reasons.

The author would like to emphasize on the basis of his experiences in China that natural and artificial noise is extremely lower than that in developed countries such as Japan. That macroscopic effects can be observed very clearly and that precursory signals can be

TABLE 1.7. List of large earthquakes in 1975 and 1976 (OIKE, 1978c).

No.	Earthquake	Province	Date	Time h	Time m	M
1	Haicheng	Liaoning	4 Feb. 1975	19	36	7.3
2	Longling	Yunnan	29 May 1976	20	23	7.5
			29 May 1976	22	00	7.6
3	Tangshan	Hebei	28 July 1976	03	42	7.8
			28 July 1976	18	45	7.1
4	Songpan-Pingwu	Sichuan	16 Aug. 1976	22	06	7.2
			22 Aug. 1976	05	49	6.7
			23 Aug. 1976	11	30	7.2
5	Yanyuan-Ninglang	Yunnan-Sichuan Border	7 Nov. 1976	02	04	6.9
			13 Dec. 1976	14	36	6.8

monitored by home-made instruments of low sensitivity may be due to the low level of noise in China, especially in provincial areas.

It is difficult to estimate the Chinese earthquake prediction program budget, in a way similar to that in western countries, because of the difference in social conditions. It is likely, however, that the level of expenditure is more or less similar to that of Japan and the U.S.A. provided Chinese prices of commodities are converted to those in the above two countries.

The Chinese began to make every effort toward absorbing foreign culture after the purge of the gang of four. In the field of earthquake prediction, it has also been planned to send missions and scholars to developed countries, to participate actively to international conferences and to receive overseas missions and scientists in China. Future development of earthquake prediction work in China will no doubt be expected in spite of the economic readjustment in recent years.

1.5 International Cooperation

The International Commission on Earthquake Prediction (ICEP), which belongs to the International Association of Seismology and the Physics of the Earth's Interior (IASPEI) of the International Union of Geodesy and Geophysics (IUGG), was created in 1967. The

ICEP convened a number of symposia on earthquake prediction whenever we had assemblies of IASPEI or IUGG. The information, especially from eastern countries, presented at these symposia was quite useful for promoting earthquake prediction study.

The ICEP activity had to be limited because it is a non-governmental organization, so that no substantial funds for carrying out actual work such as eatablishment of earthquake prediction programs in developing countries have been raised. In spite of this, bilateral exchange of missions, scientists, and instruments has been actively made between Japan, Canada, the U.S.A., and the U.S.S.R. in these years.

As a result of the increasing interest and necessity of eartquake prediction, an Intergovernmental Conference on the Assessment and Mitigation of Earthquake Risk, that was held under the auspices of UNESCO in 1976, resolved to recommend UNESCO to convene an International Symposium on Earthquake Predition.

The symposium was actually held at the UNESCO headquarters in Paris on April 2–6, 1979. The following items were covered by the symposium:

A. Approach from the side of natural science
 1. Earthquake precursors
 2. Physical processes, experiment, and theory
 3. Methods of earthquake prediction
B. Approach from the side of social science
 4. Individual and group response
 5. Economics of earthquake prediction
 6. The role of institutions
C. Interface problems
 7. Communication of predictions and warnings

One of the most outstanding features of the symposium is certainly the point that both the natural and social sides of earthquake prediction were discussed at the same time as already pointed out in the Preface and Introductory Remarks.

A panel meeting was held the week following the symposium. As a result of 4 days' discussion by about 20 eminent participants from many nations, recommendations to UNESCO were prepared. It is interesting to note that designation of international test sites, where earthquake prediction may possibly be exercised by international teams and preparation of a roster of scientists who may make earthquake prediction evaluation in response to the request of a member country, are included in the recommendations.

CHAPTER 2

OCCURRENCE PATTERN OF GREAT EARTHQUAKES—A HISTORICAL APPROACH

It is possible to see space and time pattern of occurrence of large earthquakes on the basis of historical documents concerning earthquakes in countries which have a long history of such phenomena. These countries are China, Japan, Turkey, and so on where historical documents related to earthquakes more than 2,000 years ago are sometimes available.

The Academia Sinica seems to have published three kinds of earthquake catalogue: catalogue A (ACADEMIA SINICA, SEISMOLOGICAL COMMITTEE, 1956) is the most extensive compilation in two big volumes totaling 1,653 pages. Catalogue B (ACADEMIA SINICA, GEOPHYSICAL INSTITUTE, 1971) involves brief descriptions of earthquake disaster since B.C. 1177. Meanwhile, catalogue C (ACADEMIA SINICA, GEOPHYSICAL INSTITUTE, 1976a) is just the summary of the previous ones listing the date, location, magnitude, and intensity only. The last catalogue is revised from time to time.

Important literature in Japan is as follows; Dai Nihon Jishin Shiryo (IMPERIAL EARTHQUAKE INVESTIGATION COMMITTEE, 1904, 1973). Zotei Dai Nihon Jishin Shiryo (EARTHQUAKE DISASTER PREVENTION COMMITTEE, MINISTRY OF EDUCATION, 1941–43), Nihon Jishin Shiryo (MUSHA, 1951) and Nihon Higai Jishin Soran (USAMI, 1975). Meanwhile a brief summary of historical earthquakes appears in Rikanenpyo compiled by the Tokyo Astronomical Observatory every year. Much effort is now being made searching for historical documents in Japan.

Earthquake histories in Turkey, Syria, Lebanon, and other Middle East countries are compiled by AMBRASEYS (1970).

It is inevitable that earthquake history in a country such as the U.S.A., which has an extremely short history, is incomplete. Large earthquakes in the U.S.A. during the last 250-year period are registered in the "Earthquake History of the United States" (NOAA, 1973).

The earthquake histories in Peru and Chile are compiled by LOMNITZ (1970) fairly comprehensively.

2.1 Earthquake History in China

2.1.1 Growth and decay of seismicity

Great Chinese earthquakes having a magnitude of 8 or over are picked up from the catalogue and listed in Table 2.1. The locations of these earthquakes are shown in Fig. 2.1 The worst five shocks are indicated in Table 2.2 along with brief description of the disaster. Table 2.3 summarizes the destructive earthquakes that hit the Peking (Beijing) area in history. In these tables, provinces, cities, towns, and the like are written in Roman alphabets that are used in a Chinese map published in China (MAP PRESS, 1971).

Looking at the last table, it is seen that Peking (Beijing) has been struck by strong earthquakes once in 100 years on the average. Peking (Beijing) was shaken after a long interval by the 1976 Tangshan earthquake ($M = 7.8$) and its aftershocks.

The Tancheng, Shandong Province, earthquake ($M = 8.5$, 1668) was the largest among earthquakes that occurred in south Shandon Province. The focal area of the earthquake had been very quiet during the 150-year period prior to the earthquake and, even after the earthquake, that area has been relatively quiet although a number of small shocks took place there until 1900. Since then, the area has been absolutely calm.

In view of the above-mentioned growth and decay of seismicity, it is clear that there are instances for which seismicity, which seems to have ceased, recovers after a period of a few hunderd years. No severe earthquakes occurred in the Korean Peninsula since the 17th century.

TABLE 2.1. Great earthquakes in China.

Date	Location	Latitude (N)	Longitude (E)	Magnitude	Remarks
Sept. 17, 1303	Hongtong-Zhaocheng, Shanxi Province	36.3°	111.7°	8.0	See Table 2.2.
Jan. 23, 1556	Hua Xian, Shanxi Province	34.5°	109.7°	8.0	,,
Dec. 29, 1604	Quanzhou, Fujian Province	25.0°	119.5°	8.0	Buildings were destroyed, ground cracked and water splashed from the ground over a wide area.
July 25, 1668	Tancheng-Ju Xian, Shandong Province	35.3°	118.6°	8.5	See Table 2.2.
Sept. 2, 1679	Sanhe-Pinggu, Hebei Province	40.0°	117.0°	8.0	,,
May 18, 1695	Linfen-Xiangling, Shanxi Province	36.0°	111.5°	8.0	More than 100,000 deaths.
Jan. 3, 1739	Yunchuan-Pingluo, Ningxia Province	38.9°	106.5°	8.0	More than 50,000 deaths. People were drowned in the water that came out from ground cracks.
Mar. 8, 1812	Yining, Xinjiang Province	43.7°	83.0°	7–8	Many deaths of people and cattle.
Sept. 6, 1833	Songming-Yanglin, Yunnan Province	25.2°	103.0°	8.0	6,700 deaths.
Aug. 22, 1902	Artux, Xinjiang Province	40.0°	76.5°	8.3	Gigantic ground cracks.
Dec. 23, 1906	Manas, Xinjiang Province	43.9°	85.6°	8.0	285 deaths.
June 5, 1920	Hualian, Taiwan Province	23.5°	122.0°	8.0	2 deaths.
Dec. 16, 1920	Haiyuan, Ningxia Province	36.5°	105.7°	8.5	See Table 2.2.
May 23, 1927	Gulang, Gansu Province	37.6°	102.6°	8.0	Several tens of thousand deaths.
Aug. 11, 1931	Fuyun, Xinjiang Province	47.1°	89.8°	8.0	
Aug. 15, 1950	Zayü, Xizang Province	28.4°	96.7°	8.5	Tibet-Assam earthquake.
Nov. 18, 1951	Damxung, Xizang Province	31.1°	91.4°	8.0	
Jan. 25, 1972	Xingang, Taiwan Province	23.0°	122.3°	8.0	1 death.
Feb. 6, 1973	Luhuo, Sichuan Province	31.3°	100.9°	7.9	

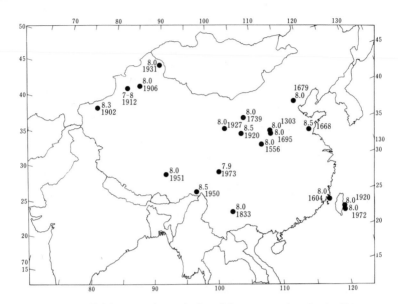

FIG. 2.1. Epicenters and magnitudes of the great earthquakes in China.

It is known, however, that many destructive earthquakes had occurred there before the 17th century. It cannot be ruled out that seismicity may revive in the area some time in the future. This is also the case for the Kyoto area, Japan, where we had no earthquakes of large magnitude for a few hundred years. Even in the eastern U.S.A. where very few large earthquakes have been recorded in the relatively short history, nobody guarantees that a large earthquake would not occur there in the future.

2.1.2 Migration of seismic activity

If one looks at the "Distribution Map of Strong Earthquakes in China" (ACADEMIA SINICA, GEOPHYSICAL INSTITUTE, 1976b), it is seen that large earthquakes have been occurring along 3 lines running from northeast to southwest through Liaoning, Hebei, and Shandong Provinces. Such a tendency can also be seen, though not quite clear, even in the epicenter map of great earthquakes shown in Fig. 2.1. These lines belong to a fault system which has been active throughout the Quaternary period.

The seismic activity connected to the fault system had been low

TABLE 2.2. Five greatest earthquakes in China.

Date	Location	Latitude (N)	Longitude (E)	Magnitude	Remarks
Sept. 17, 1303	Hongtong-Zhaocheng, Shanxi Province	36.3°	117.8°	8	
Jan. 23, 1556	Hua Xian, Shaanxi Province	34.5°	109.7°	8	820,000 deaths.
July 25, 1668	Tancheng-Ju Xian, Shangong Province	35.3°	118.6°	8.5	Seismicity is low over a long period before and after the earthquake.
Sept. 2, 1679	Sanhe-Pinggu, Hebei Province	40.0°	117.0°	8	Violent shock in Peking.
Dec. 16, 1920	Haiyuan, Ningxia Province	36.5°	105.7°	8.5	180,000 deaths.

TABLE 2.3. Destructive earthquakes in the Peking (Beijing) area.

Date	Location	Latitude (N)	Longitude (E)	Magnitude	Remarks
1057	Gu'an, Hebei Province	39.5°	116.3°	6.8	Several tens of thousand deaths.
Sept. 8, 1337	Huailai, Hebei Province	40.4°	115.7°	6.5	Violent shock in Peking.
Jan. 29, 1484	Juyongguan, Peking	40.4°	116.1°	6.8	Destruction in North Peking.
Oct. 22, 1536	Tong Xian, Hebei Province	39.8°	116.8°	6.0	Destruction in Tongzhou.
Feb. 3, 1658	Laishui, Hebei Province	39.4°	115.7°	6.0	Many deaths.
Apr. 16, 1665	Tong Xian, Hebei Province	39.9°	116.7°	6.5	Countless houses were destroyed in Peking.
Sept. 2, 1679	Sanhe-Pinggu, Hebei Province	40.0°	117.0°	8.0	See Table 2.2.
Sept. 30, 1730	Western suburb, Peking	40.0°	116.2°	6.5	More than 10,000 houses were destroyed in and around Peking.

in this century unitil the Xingtai earthquakes ($M = 6.8$, 7.2) occurred in 1966. Since then, the activity has migrated in a northeast direction as can be seen in Fig. 2.2 culminating in the 1975 Haicheng earthquake ($M = 7.3$). It is not entirely clear why such a migration does take place. SCHOLZ (1977) introduced a concept called the "deformation front." But nothing is made clear physically even though such a terminology is introduced.

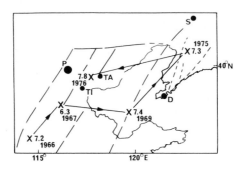

FIG. 2.2.   Migration of large earthquake epicenters in Hebei and Liaoning Provinces, China. D: Dalian, P: Peking (Beijing), S: Shenyang, TA: Tangshan, TI: Tianjin.

The migration to the northeast direction tended to change its direction resulting in the Tangshan earthquake in 1976. According to OIKE (1978a), it is said that the Chinese noted the tendency of westward migration of seismic activity after the 1668 Tancheng, Shandong Province, earthquake ($M = 8.5$). It must have been anticipated, therefore, that, should the migration tendency in the 17th century be assumed to repeat, the Peking (Beijing)-Tianjin-Tangshan area might be the next target after the Haicheng earthquake. As Peking (Beijing) was outside of the hazard area of the Tangshan earthquake, there is a suggestion that Peking (Beijing) could be the seat of the next large earthquake.

It is pointed out that a large earthquake is apt to occur at the intersection between an active fault and a geological structural line. Pattern recognition studies by GELFAND et al. (1972, 1976) support on the basis of the analyses of earthquakes in Middle Asia and California such a view that earthquakes occur at intersections of lineaments.

According to OIKE (1978a), there were two instances for which epicenters moved from Peking (Beijing) to Xi'an along the Fen River in Chinese history. The first migration took place during 516–1568, and the most disastrous Hua Xian, Shaanxi Province, earthquake ($M=8$, 1556; See Table 2.2) occurred towards the end of the migration period. The second migration was observed during 1626–1815.

Sichuan and Yunnan Provinces are situated on an earthquake belt, and so struck by large earthquakes from time to time. During the 1978 visit to China, the author was told by the members of the Sichuan Seismological Bureau that they observed a marked tendency of alternate occurrence of large earthquakes between both the provinces as shown in Fig. 2.3. For example, the Longling, Yunnan Province, earthquakes ($M=7.5, 7.6$), that occurred on May 29, 1976, were followed by the Songpan-Pingwu, Sichuan Province, earthquakes ($M=7.2$, August 16; $M=6.7$, August 22; $M=7.2$, August 23, 1976). Subsequently, the Yanyuan-Ninglang earthquakes ($M=6.9$, November 7; $M=6.8$, December 13) occurred at the Sichuan-Yunnan border in the same year. The tendency of alternate occurrence of Yunnan-Sichuan earthquakes is so distinct that a long-term prediction of earthquake occurrence is actually made although no physical mechanism of such a tendency is known.

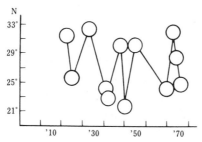

Fig. 2.3. Alternate occurrences of large earthquakes of magnitude 7 or over in Sinchuan and Yunnan Provinces.

That epicenters of great earthquakes tend to migrate in a certain direction has been noticed in and around Japan, the Circum-Pacific area and Turkish Anatolia (Mogi, 1968a, b; Ambraseys, 1970). In the Turkish instance, the westward drift was very clear after the Erzincan earthquake ($M=8.0$, 1939).

In spite of the marked tendency of epicentral migration in some particular cases, it appears to the author that we cannot entirely rely on such migration tendency in paractical earthquake prediction although it is natural to pay attention to such a tendency, if any, for synthesized judgment.

2.2   Recurrence of Great Earthquakes at Subduction Zones

No clear-cut theory about the mechanism of large intra-continental earthquakes in China, Turkey, and so on has been presented even if we sometimes observe a very regular migration tendency of epicenter. Although it has been suggested that the stress field that generates those Chinese earthquakes is primarily due to the northward motion of the Indian plate and the westward motion of the Pacific plate relative to the Eurasia plate, nothing in detail has as yet been brought to light.

In contrast to such a situation of the interpretation of intra-continental earthquakes, some regularities of generation mechanism of large earthquakes at subduction zones can be put forward on the basis of the present-day plate tectonics. The recurrence tendency of great earthquakes as will be discussed in the following may possibly be interpreted physically by the plate tectonics.

2.2.1   Great earthquakes in Japan

Figure 2.4 shows the distribution of destructive earthquakes that occurred in and around Japan since 1600. Epicenters of earthquakes for which $M \geq 7.9$ are shown with solid circles. These earthquakes are hereafter called the great earthquakes, and are listed in Table 2.4. Magnitude values estimated for earthquakes before 1920 are not accurate. Therefore, revised magnitudes are sometimes shown with parentheses in the table.

The profound feature of Fig. 2.4 is certainly the point that almost all of great earthquakes occur along or off the Pacific coast except the 1891 Nobi earthquake on Honshu Island.

According to the current idea of modern plate tectonics, the Pacific plate, that is born at the East Pacific Rise and proceeds in a northwest direction, tends to subside into the earth's interior at the

FIG. 2.4.   Destructive earthquakes in Japan during 1600–1964 including the Tokachi
Oki earthquake ($M = 7.9$, 1968). Earthquakes for $M \geq 7.9$ are indicated with solid
circles.

Japan-Kurile-Kamchatka-Aleutian trenches pulling down and com-
pressing the continental plate there. When the deformation due to the
plate motion reaches a critical value, rupture takes place at the
interface between the land and sea plates. In that case the land plate,
which has been pulled down and compressed horizontally, rebounds
almost instantaneously giving rise to upward and seaward displacement of land.

At the time of rupture, the strain energy accumulated in the
earth's crust by that time is radiated as that of seismic waves. Because
of the sudden change in sea-bottom topography in association with
the rebound, a large-scale disturbance of sea water suddenly occurs
producing tsunami waves.

The above-mentioned process of subduction—rupture—rebound can account for the focal mechanism of great earthquakes
at subduction zones along with the crustal movement of reverse
fault type associated with these earthquakes. It has therefore become widely accepted in these days that the said process provides a highly plausible physical model of what is going on at the

TABLE 2.4. Great earthquakes in Japan.

Date	Focal area	Latitude (N)	Longitude (E)	Magnitude	Remarks
Hokkaido-Kurile					
Apr. 25, 1843	Off Kushiro	42.0°	146.0°	8.4	
Mar. 22, 1894	Off Nemuro	42.4	146.3	7.9 (7.4)	
Sept. 8, 1918	Off Urup Is.	45.7	151.8	7.9 (7.4)	
Mar. 4, 1952	Off Tokachi	42.2	143.9	8.1	
Nov. 7, 1958	Off Itrup Is.	44.3	148.5	8.0	
Oct. 13, 1963	Off Itrup Is.	43.8	150.0	8.1	
May 16, 1968	Off Tokachi	40.7	143.6	7.9	
Off Sanriku (Off Pacific coast of Northeast Japan)					
July 13, 869	Off Sanriku	38.5	143.8	8.6	
Dec. 2, 1611	Off Sanriku	38.2	148.8	8.1	
Apr. 13, 1677	Off Sanriku	38.7	144.0	8.1	
Mar. 3, 1933	Off Sanriku	39.1	144.7	8.3	
Kanto (Tokyo area)					
818	Sagami Trough (?)	35.2	139.3	7.9	It is suspected that the epicenter is located in the middle of Kanto plain.
Feb. 3, 1605	Sagami Trough	34.3	140.4	7.9	
Dec. 31, 1703	Sagami Trough	34.7	139.8	8.2	
Sept. 1, 1923	Sagami Trough	35.2	139.3	7.9	

TABLE 2.4 (continued)

Date	Focal area	Latitude (N)	Longitude (E)	Magnitude	Remarks
Middle Honshu (Middle part of the main island of Japan)					
June 5, 745	Mino	35.5°	136.6°	7.9	
Jan. 18, 1586	Hida	36.0	136.8	7.9	
Oct. 28, 1891	Gifu, Aichi	35.6	136.6	8.4 (7.9)	
Off Tokai (Off Pacific coast of Central Japan)					
Dec. 17, 1096	Kumano-nada Sea	34.2	137.3	8.4	
Sept. 20, 1498	Enshu-nada Sea	34.1	138.2	8.6	
Dec. 23, 1854	Enshu-nada Sea	34.1	137.8	8.4	
Dec. 7, 1944	Kumano-nada Sea	33.7	136.2	8.0	The shock is followed by the Nankai earthquake after 32 hours.
Off Nankai (Off Kii Peninsula and Shikoku Is.)					
Nov. 29, 684	Off Nankai	32.5	134.0	8.4	
Aug. 26, 887	Off Nankai	33.0	135.3	8.6	Pair shock (?)
Feb. 22, 1099	Off Nankai	33.0	135.5	8.0	
Aug. 3, 1361	Off Nankai	33.0	135.0	8.4	
Feb. 3, 1605	Off Nankai	33.0	134.9	7.9	Pair shock.
Oct. 28, 1707	Off Nankai	33.2	135.9	8.4	Pair shock.
Dec. 24, 1854	Off Nankai	33.2	135.6	8.4	Pair shock.
Dec. 21, 1946	Off Shikoku and Kii	33.0	135.6	8.1	
South of Kyushu					
June 24, 1901	Near Amami-Oshima Is.	28.3	129.3	7.9 (7.4)	
Nov. 10, 1909	Hyuga-nada Sea	32.1	133.1	7.9 (7.4)	
June 15, 1911	Near Kikaiga-shima Is.	28.0	130.0	8.2 (7.7)	

focal region of a great earthquake at subduction zones.

A more detailed configuration of plates in Central Japan is shown in Fig. 2.5 in which the two deep sea canyons called the Sagami and Nankai troughs are shown. There is evidence that the Philippine Sea plate is moving in a northwest direction. KANAMORI (1972a), who studied the focal mechanism of the Tonankai ($M = 8.0$, 1944) and Nankai ($M = 8.1$, 1946) earthquakes on the basis of world-wide seismograms, indicated that the subduction—rupture—rebound model at the Nankai trough may well account for the seismograms observed.

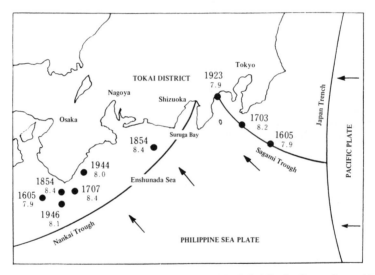

FIG. 2.5.   The plates, troughs and trench around the Tokai district, Japan along with the epicenters of great earthquakes since the 17th century.

The space-time distribution of great earthquakes in the Nankai and Tokai zone probably related to the subduction at the Nankai trough is shown in Fig. 2.6. As can be seen in the figure, it is really noticeable that great earthquakes have been originating from almost the same area with a return period of 120 years or so. According to the data since 1600, the mean return period and its standard deviation of recurrence of great earthquakes in the Nankai-Tokai zone are estimated as 117 and 35.0 years, respectively.

Much of recurrence tendency of great earthquakes in the

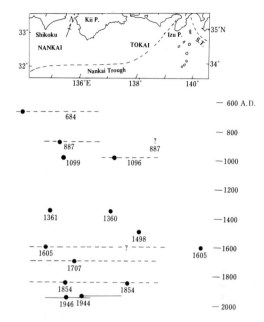

FIG. 2.6.   Space-time distribution of great earthquakes in the Nankai-Tokai zone. The rupture zones are indicated with solid-line segments. The segments shown with a dashed line are uncertain because they are estimated from classical documents.

Hokkaido-Kurile zone has been discussed by UTSU (1972). The space-time distribution of great earthquakes there are shown in Fig. 2.7 in which blocks A, B, C,... as defined by Utsu correspond to the aftershock areas of recent earthquakes. Almost no overlap between neighboring aftershock areas has been noticed. The mean return period and its standard deviation is estimated as 85.3 and 24.6 years, respectively, for the zone.

In contrast to the above two zones, no regular tendency of recurrence can be seen along the Sanriku-Boso zone which runs off the Pacific coast of northeastern Japan. KANAMORI (1972b) pointed out the physical reason that accounts for the said tendency. He also proved that the great Sanriku earthquake ($M = 8.3$, 1933) was caused by a normal fault instead of a reverse fault which is the natural consequence of the subduction—rupture—rebound process.

No clear tendency of recurrence can be seen at the Sagami trough where the sea and land plates pass each other although great

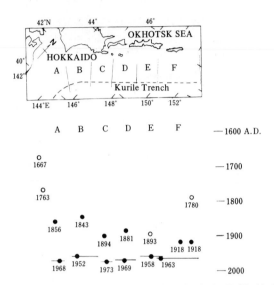

Fig. 2.7. Space-time distribution of great earthquakes in the Hokkaido-Kurile zone. Uncertain locations of epicenter are shown with open circles. Other legends are the same as those in Fig. 2.6.

earthquakes such as the Kanto earthquake ($M = 7.9$, 1923) occur there from time to time.

2.2.2 Great earthquakes around the Pacific Ocean

Figures 2.8–2.11 show the space-time pattern of occurrence of great earthquakes respectively for the Kamchatka, Aleutian-Alaska, Central America, and South America zones. In Tables 2.5–2.8 are listed these earthquakes. The data are taken from FEDOTOV et al., (1970, 1972), KELLEHER (1970, 1972), KELLEHER and SAVINO (1975), KELLEHER et al. (1973, 1974), and SYKES (1971).

Figure 2.12 is the histograms of return period for recurrence of great earthquakes at the respective subduction zones around the Pacific Ocean (RIKITAKE, 1976b). The important feature of the histograms in Fig. 2.12 is certainly the point that the maximum frequency differs considerably each other from zone to zone.

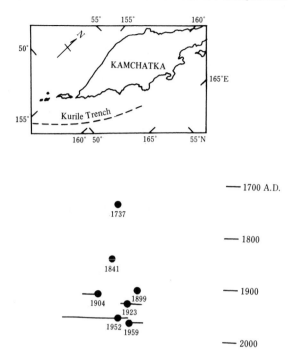

FIG. 2.8   Space-time distribution of great earthquakes in the Kamchatka zone.

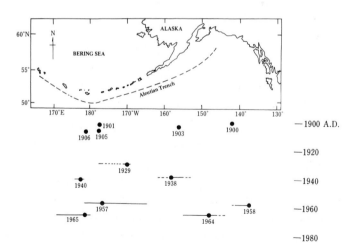

FIG. 2.9.   Space-time distribution of great earthquakes in the Aleutian-Alaska zone.

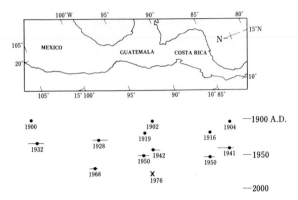

Fig. 2.10. Space-time distribution of great earthquakes in the Central America zone.

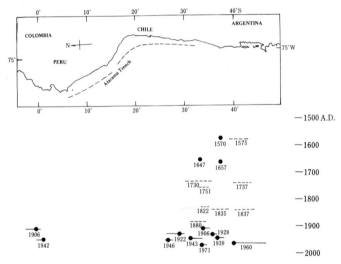

Fig. 2.11. Space-time distribution of great earthquakes in the northern and southern South America zones.

2.2.3 Weibull distribution analysis of return period of great earthquakes

Rikitake (1976b) analysed the recurrence data by making use of a Weibull distribution analysis which was first introduced to the earthquake prediction problem by Hagiwara (1974).

Let us denote the probability of an earthquake occurring between t and $t + \Delta t$ by $\lambda(t)$ on the condition that no earthquake occurred

TABLE 2.5. Great earthquakes in Kamchatka.

Date	Focal area	Latitude (N)	Longitude (E)	Magnitude	Remarks
Nov. 23, 1899		53.0°	159.0°	7.9	
June 25, 1904		52.0	159.0	8.3	Aftershocks of magnitude 8.1 and 7.9.
Jan. 30, 1917		56.5	163.0	8.1	
Feb. 3, 1923		54.0	161.0	8.4	Tsunamis of 5–6 m in wave height hit east coast of Kamchatka. Six deaths by tsunamis at Hilo, Hawaii Is.
Nov. 4, 1952		52.8	159.5	8.4	Large tsunamis. Wave height reached 18.4 m at Paramushir Is.
May 4, 1959		52.5	159.5	8.3	1 dead, 13 injured.

TABLE 2.6. Great earthquakes in Aleutian and Alaska.

Date	Focal area	Latitude (N)	Longitude	Magnitude	Remarks
Oct. 9, 1900	Alaska Peninsula	60.0°	142.0°W	8.2	
Dec. 31, 1901	Andreanof Is.	52.0	177.0 W	7.8	
June 2, 1903	Alaska Peninsula	57.0	156.0 W	8.3	
Feb. 14, 1905	Andreanof Is.	53.0	178.0 W	7.9	
Aug. 17, 1906	Rat Is.	51.0	179.0 E	8.3	
Mar. 7, 1929	Andreanof Is.	51.0	170.0 W	8.6	Tsunamis in Hawaii. $M=8.1$ according to Rikanenpyo.
Nov. 11, 1938	Alaska Peninsula	55.5	158.0 W	8.7	Small tsunamis in Hawaii. $M=8.3$ according to Rikanenpyo.
July 14, 1940	Rat Is.	51.8	177.5 E	7.8	
Mar. 9, 1957	Andreanof Is.	51.3	175.8 W	8.3	
July 10, 1958	Southern Alaska	58.6	137.1 W	7.9	6 dead. Rockslide excited huge waves of 1,740 feet in height in Lituya Bay.
Mar. 28, 1964	Southern Alaska	61.1	147.6 W	8.4	Good Friday earthquake. 114 dead or missing, 12 dead by tsunamis.
Feb. 4, 1965	Rat Is.	51.3	178.6 E	7.8	Tsunamis in Attu Is. of 2.4 m in height.

TABLE 2.7. Great earthquakes in Central America.

Date	Focal area	Latitude (N)	Longitude (W)	Magnitude	Remarks
Jan. 20, 1900	Jalisco coast, Mexico	20.0°	105.0°	8.2	2,000 dead.
Apr. 19, 1902	Guatemala	14.0	91.0	8.2	
Sept. 23, 1902	Chiapas, Mexico	16.0	93.0	8.4	
Jan. 14, 1903	Off Oaxaca, Mexico	15.0	98.0	8.2	
Apr. 15, 1907	Guerrero, Mexico	17.0	100.0	8.1	28 dead. Tsunamis.
June 3, 1932	Jalisco coast, Mexico	19.5	104.3	8.1	60 dead. 400 injured. Tsunamis.
Aug. 6, 1942	Guatemala	14.0	91.0	7.9	

TABLE 2.8. Great earthquakes in South America.

Date	Focal area	Latitude	Longitude (W)	Magnitude	Remarks
Ecuador-Colombia area					
Jan. 31, 1906	Off Ecuador	1.0°N	81.5°	8.6	Tsunamis at Tumaco, Colombia.
May 14, 1942	Off Ecuador	0.8 S	81.5	7.9	Some destruction.
Atacama area, Chile					
Nov. 24, 1604	Off Arica	—	—	8.3–8.5	Large tsunamis.
Apr. 3, 1819	Off Copiapo	—	—	8.3–8.5	Large tsunamis.
Aug. 13, 1868	Off Arica	—	—	8.5	Large tsunamis.
May 9, 1877	Off Pisagua	—	—	8–8.5	Large tsunamis.
Nov. 11, 1922	Off Central Chile	28.5 S	70.0	8.3	Tsunamis of 7 m in height at Caldera.
Aug. 2, 1946	Near coast of northern Chile	26.5 S	70.5	7.5	
Santiago-Valparaiso area					
May 13, 1647	Between Santiago and Valparaiso	—	—	8.5	
July 8, 1730	Off Valparaiso	—	—	8.8	Large tsunamis.
Nov. 19, 1822	Near coast of Valparaiso	—	—	8.5	Land deformation.

TABLE 2.8 (continued)

Date	Focal area	Latitude	Longitude (W)	Magnitude	Remarks
Aug. 15, 1880	Near coast of Illapel	—	—	7.5–8	Small tsunamis, land deformation.
Aug. 17, 1906	Near coast of Valparaiso	33.0°S	72.0°	8.4	Small tsunamis.
Apr. 6, 1943	Near coast of Illapel	30.8 S	72.0	7.9	Tsunamis of 1.2 m in wave height.
July 9, 1971	Near coast of Valparaiso	32.5 S	71.2	7.5	83 dead, 447 injured.
Concepcion area					
Feb. 8, 1570	Off Concepcion	—	—	8–8.5	Large tsunamis.
Mar. 15, 1657	Off Concepcion	—	—	8	Large tsunamis.
May 25, 1751	Off Concepcion	—	—	8.5	Large tsunamis.
Feb. 20, 1835	Off Concepcion	—	—	8–8.3	Large tsunamis.
Dec. 1, 1928	Coast of Talca	35.0 S	72.0	8.0	Small tsunamis, land deformation, 218 dead.
Jan. 25, 1939	Near Chillan	36.3 S	72.3	7.8	3,000 dead.
Valdivia area					
Dec. 16, 1575	Near Valdivia	—	—	8.5	Tsunamis.
Dec. 24, 1737	Off Valdivia	—	—	7.5–8	Tsunamis.
Nov. 7, 1837	Off Valdivia	—	—	8	
May 22, 1960	Off Valdivia	39.5 S	74.5	8.5	Tsunami destruction even in Japan.

68

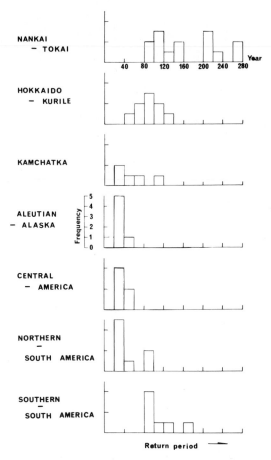

FIG. 2.12. Histograms of return period of great earthquakes at subduction zones.

before t. It is assumed that $\lambda(t)$ is expressed by a Weibull distribution such as

$$\lambda(t) = Kt^m \qquad (2.1)$$

in which parameters K and m are constants and $K > 0$ and $m > -1$.

The cumulative probability of an earthquake occurring between 0 and t is denoted by $F(t)$, and the reliability $R(t)$ is defined by

$$R(t) = 1 - F(t) \qquad (2.2)$$

On the other hand, $R(t)$ can be calculated as

$$R(t) = \exp\left[-\int_0^t \lambda(t)dt \right] = \exp\left(-\frac{Kt^{m+1}}{m+1} \right) \tag{2.3}$$

As the probability density of earthquake occurrence becomes

$$f(t) = -\frac{dR}{dt} = Kt^m \exp\left(-\frac{Kt^{m+1}}{m+1} \right) \tag{2.4}$$

The mean return period is calculated as

$$E[t] = \int_0^\infty tf(t)dt = \left(\frac{K}{m+1} \right)^{-1/(m+1)} \Gamma\left(\frac{m+2}{m+1} \right) \tag{2.5}$$

where Γ is a Gamma function.

We further have

$$E[t^2] = \int_0^\infty t^2 f(t)dt = \left(\frac{K}{m+1} \right)^{-2/(m+1)} \Gamma\left(\frac{m+3}{m+1} \right) \tag{2.6}$$

so that the standard deviation of mean return period is given as

$$(E[t^2] - E^2[t])^{1/2} = E[t]\left[\Gamma\left(\frac{m+3}{m+1} \right) - \Gamma^2\left(\frac{m+2}{m+1} \right) \right]^{1/2} \bigg/ \Gamma\left(\frac{m+2}{m+1} \right) \tag{2.7}$$

When we take a double logarithm of $1/R(t)$, we obtain

$$\log_e \log_e (1/R) = \log_e [K/(m+1)] + (m+1) \log_e t \tag{2.8}$$

which is useful for determining K and m from actually observed data by means of the least-squares method.

In practice, we count the frequency of return period (n_i) for each time range Δt suitably chosen. The probability of a return period falling in a range between $i\Delta t$ and $(i+1)\Delta t$ ($i=0, 1, 2, 3, \ldots$) is obtained as n_i/N for which N is the total number of data. The cumulative probability is then obtained as

$$F = \sum_{i=0}^i n_i/N \tag{2.9}$$

so that R can readily be obtained from Eq. (2.2).

With R thus calculated, $\log_e \log_e (1/R)$ versus $\log_e t$ plots are made by adopting an appropriate time interval. As a result of the best straight-line fitting for the plotted data, we can determine $\log_e [K/(m+1)]$ and $m+1$ which are the two constants specifying the straight line. Two parameters K and m are then calculated.

Once K and m are obtained, mean return period and its standerd deviation are readily calculated from Eqs. (2.5) and (2.7). The mean return periods are given in Table 2.9 along with the standard deviations for the respective zones. As the reliability R can be calculated from Eq. (2.3), the cumulative probability F for earthquake occurrence is obtained from Eq. (2.2). The cumulative probabilities thus estimated for the respective zones are shown in Fig. 2.13.

TABLE 2.9.   Mean return period and its standard deviation for each zone.

Zone	Mean return period (year)	Standard deviation (year)
Nankai-Tokai	117.0	35.0
Hokkaido-Kurile	85.3	24.6
Aleutian-Alaska	27.2	8.9
Central America	34.5	3.6
Northern South America	46.3	30.0
Southern South America	100	22.5

Looking at the curves in Fig. 2.13, it is remarkable that the increasing rate of probability is considerably different from zone to zone. Should the space-time pattern of occurrence of great earthquakes be unchangeable, the curves shown in Fig. 2.13 may give some clue to estimating the recurrence probability of a great earthquake for each zone even though no physical mechanism as to why great earthquakes tend to recur in a manner as argued here is known.

The steep rise of the probability curve for the Central America zone is really remarkable, so that the estimate of the year in which a great earthquake recurs may be made semi-deterministically with an accuracy of several years. Actually, the 1976 Guatemala earthquake as denoted by a cross in Fig. 2.10 occurred exactly 34 years after the last quake.

It is important that a probability of earthquake occurrence can

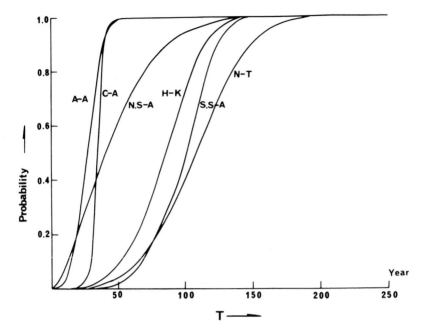

Fig. 2.13. Cumulative probability of a great earthquake recurring at the respective zones. N-T: Nankai-Tokai zone, H-K: Hokkaido-Kurile zone, A-A: Aleutian-Alaska zone, C-A: Central America zone, N.S-A: northern South America zone, S.S-A: southern South America zone.

sometimes be estimated from simple statistics of recurrences of great earthquakes as can be see in this subsection. It should be borne in mind that any statistics should be supported by some physically acceptable mechanism. The subduction—rupture—rebound process based on the plate tectonics provides the background of the statistics made here. Scatterings of return period may be interpleted as fluctuations of the above process.

2.3 Seismic Gap

If one admits that the physical mechanism for occurrence of great earthquakes as discussed in the last section is true, it is naturally expected to have an area, where no large earthquake occurs over a fairly long period, preceding the occurrence of the next large earthquake because some period is required for strain accumulation. Such an aseismic area is called the seismic gap of the first kind (MOGI, 1978). The concept of seismic gap has been developed by FEDOTOV (1965), MOGI (1968b), SYKES (1971) and others. It is well known that the focal area of the Kanto earthquake ($M = 7.9$, 1923) is at present a seismic gap (SHIMAZAKI, 1971), that the Nemuro-Hanto Oki earthquake ($M = 7.4$, 1973) actually occurred in a seismic gap off Nemuro, Hokkaido (UTSU, 1972), and that there is a remarkable seismic gap in the Enshunada Sea off the Tokai area where we expect a large earthquake to occur sooner or later (SEKIYA and TOKUNAGA, 1974).

USGS published the proceedings of a symposium on seismic gap (EVERNDEN et al., 1978) in which MCCANN et al.(1978) made a comprehensive review of existing gaps. It is interesting to look at a table in which MCCANN et al. (1978) listed successful forecasts based on a seismic gap. The table is reproduced in Table 2.10.

TABLE 2.10. Successful forecasts based on a seismic gap (MCCANN et al., 1978).

Forecast		Earthquake		
Investigator	Year published	Location	Year	Magnitude*
Fedotov	1965	Near Hokkaido, Japan (Tokachi-oki)	1968	7.9
Fedotov	1965	Near southern Kurile Islands	1969	7.8
Fedotov	1965	Near central Kamchatka	1971	7.3
Fedotov	1965	Near Hokkaido, Japan		
Mogi	1968	(Nemuro-oki)	1973	7.4
Sykes	1971	Near Sitka southeastern Alaska	1972	7.2
Kelleher	1972	Near Lima, Peru	1974	7.5
Kelleher & others	1973	Near Colima, Mexico	1973	7.3

* Magnitudes are taken from the Rikanenpyo.

MOGI (1978) further defined the seismic gap of the second kind. He pointed out a tendency that a decrease in the activity of small to moderate earthquakes takes place in an area prior to occurrence of a large earthquake. As gaps of this kind may be some kind of earthquake precursor, mention will be made of in Subsection 4.4.3.

CHAPTER 3

MONITORING OF CRUSTAL STRAIN

The subduction—rupture—rebound process as stated in the preceding chapter suggests that we may be able to say something about occurrence of a great earthquake provided the extent of accumulation of crustal strain can be monitored. It would be more realistic to talk about occurrence of earthquake on the basis of the actually-monitored crustal strain to be revealed by repetition of geodetic surveys than to say something only based on historical records.

3.1 Accumulation Rate of Crustal Strain

3.1.1 Sagami Bay and its surroundings

Japan is covered by a dense network of geodetic stations as is demonstrated in Figs. 1.1 and 1.2. The South Kanto area, an area to the south of Tokyo, including the Sagami Bay and its surroundings has been the target of frequent and precise geodetic surveys in recent years.

That the triangulation stations in the South Kanto area had displaced by several meters in a southeast direction in association with the 1923 Kanto earthquake ($M = 7.9$, 1923) was confirmed by comparing the triangulation survey in 1925 to that in 1891 (MUTO, 1932; RIKITAKE, 1976a, Fig. 5.4). A close look at the displacement pattern indicates, however, that the stations on Izu Peninsula and Izu-Oshima Island moved in an opposite direction. Such a crustal deformation may well be explained by assuming that the land plate, that had been dragged by the north-westward motion of the Philippine Sea plate, became free by the rupture at the Sagami trough and rebounded to a southeast direction.

ANDO (1971, 1974) estimated that the fault plane associated with the Kanto earthquake has a length and a width respectively amounting to 85 and 55 km. The fault trace, that is the intersection of the fault plane and the earth's surface, agrees with the trough, while the fault plane dips north-eastward with an angle of 30 degrees. The displacement in the direction of fault strike was estimated as 6 m, while a 3 m displacement took place in a direction perpendicular to the strike. Such a source model seems to agree approximately with that obtained from the analyses of world-wide seismograms (KANAMORI and MIYAMURA, 1970; KANAMORI, 1971a).

In the beginning of the 1970's, GSI conducted a trilateration survey over the South Kanto area, and the result is compared to that of a triangulation survey soon after the Kanto earthquake. It is thus disclosed that the distance between the two stations on Izu-Oshima Island and Izu Peninsula was shortened by 100 cm, while those between the stations on Izu-Oshima Island and Boso and Miura Peninsulas were lengthened by several tens of centimeters as can be seen in Fig. 3.1 (GEOGRAPHICAL SURVEY INSTITUTE, 1972).

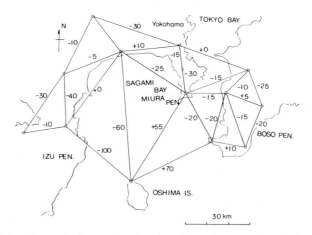

FIG. 3.1. Changes in distance in units of centimeters between triangulation stations in the South Kanto area during 1925–71 (GEOGRAPHICAL SURVEY INSTITUTE, 1972).

The crustal movement as indicated in Fig. 3.1 can be accounted for by a north-westward motion of the Philippine Sea plate provided that the fault appearing at the Sagami trough at the time of the Kanto earthquake has been locked during the post-earthquake period and so the land plate has been subjected to the drag by the sea plate. It is possible from the results shown in Fig. 3.1 to estimate horizontal strains accumulated in any triangles during the two surveys. Such an estimate reveals that one-third of the crustal movement that occurred in association with the earthquake has recovered during the 46-year period.

The accumulation rate of maximum shearing strain averaged for a number of triangles covering Sagami Bay is estimated as $0.0568 \times 10^{-5}/\mathrm{yr}$.

3.1.2 Tokai area

As has been discussed in Chapter 2, no great earthquake occurred in the Tokai area (where the mean return period of recurrence of great earthquakes is estimated as 117 years) during the last 127 years since the 1854 earthquake of magnitude 8.4. Local people simply fear that another great earthquake is about due because it is the right time to expect one, judging from the occurrence pattern in history.

It is the aim of this subsection to bring out the present state of crustal strain in the Tokai area hoping to provide data which are useful for estimating the probability of earthquake occurrence.

First of all, Fig. 3.2 (GEOGRAPHICAL SURVEY INSTITUTE, 1974c) indicates the changes in distance between neighboring first-order triangulation stations along a line crossing the central Japan in a northwest direction during 1890–1973. It is clear from the figure that all the neighboring distances shortened during the period. The shortened length in total amounts to about 3 m for the entire distance of some 200 km. Such a contraction of the earth's crust may well be understood by the compression due to the north-westward motion of the Philipine Sea plate.

According to GSI's surveys in 1884 and 1973, the distances between a number of triangular stations in an area surrounding the Suruga Bay changed in a manner shown in Fig. 3.3 in which

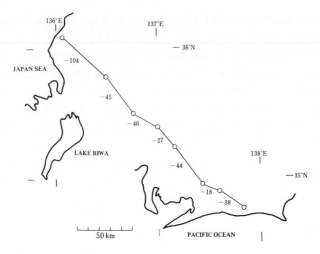

FIG. 3.2.   Changes in distance in units of centimeters between adjoining triangulation stations across Honshu in Central Japan during 1890–1973 (GEOGRAPHICAL SURVEY INSTITUTE, 1974c).

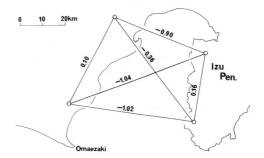

FIG. 3.3.   Changes in distance in units of meters between triangular stations in the vicinity of the Suruga Bay during 1884–1973 (GEOGRAPHICAL SURVEY INSTITUTE, unpublished, 1979).

outstanding shortening of lines crossing the bay is brought to light (GEOGRAPHICAL SURVEY INSTITUTE, unpublished, 1979). GSI (1977) has already published similar changes during 1931–73. It may be approximately said that the accumulation rate of crustal strain has been more or less the same over the 90-year period. The accumulation rate of maximum shearing strain as deduced from the 1931–73 result for triangles covering the bay is estimated as 0.043×10^{-5}/yr on the average. Figure 3.3 also demonstrates the crustal contraction due to the plate motion.

As distance measurement has frequently been conducted over the bay area under the intensified program on earthquake prediction in recent years, it is brought out that the said contraction is now in progress with a rather increased speed as can be seen in Fig. 3.4 (GEOGRAPHICAL SURVEY INSTITUTE, 1979c)

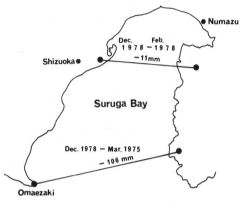

FIG. 3.4. Changes in distance in units of millimeters between triangulation stations in the vicinity of the Suruga Bay during the periods indicated in the figure (GEOGRAPHICAL SURVEY INSTITUTE, 1979c).

The west coast of Suruga Bay has been subsiding ever since levelling survey was introduced in Japan. The overall change in level in the Tokai area during 1900–73 is shown in Fig. 3.5 (GEOGRAPHICAL SURVEY INSTITUTE, unpublished, 1979) which may roughly be compatible with the supposed subduction of the Philippine Sea plate at the Nankai-Suruga trough.

Figure 3.6, which shows the change in height of a few bench marks in these 10–30 years, also indicates a clear tendency of subsidence that is prevailing even at present with a rather increasing rate (GEOGRAPHICAL SURVEY INSTITUTE, 1979a, 1980a, b, 1981). The cause of fluctuations at Bench March 2595 are not known.

Quite independent evidence of the above-mentioned land subsidence can be effected by analysing changes in the sea level at tide-

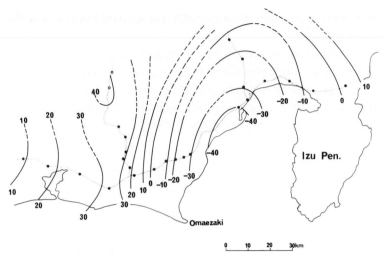

FIG. 3.5. Changes in height in units of centimeters in the Tokai area during 1900–73 as revealed by levelling surveys (GEOGRAPHICAL SURVEY INSTITUTE, unpublished, 1979).

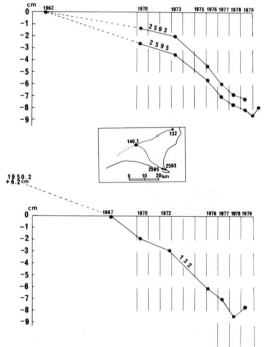

FIG. 3.6. Secular subsidence of the levelling bench marks around Omaezaki in recent years (GEOGRAPHICAL SURVEY INSTITUTE, 1979a, 1980a, b, 1981).

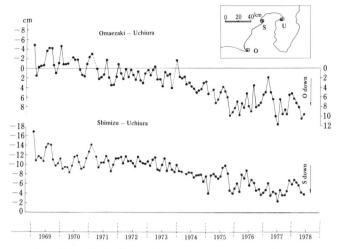

FIG. 3.7.   Locations of the three tide-gauge stations around the Suruga Bay: Omaezaki (O), Shimizu (S) and Uchiura (U). The upper and lower graphs show the differences in monthly mean sea level between Omaezaki and Uchiura and Shimizu and Uchiura, respectively. The arrows indicate land subsidence at Omaezaki and Shimizu relative to Uchiura (GEOGRAPHICAL SURVEY INSTITUTE, unpublished, 1978).

gauge stations around the Suruga Bay. The upper and lower graphs in Fig. 3.7 show the monthly means of sea-level change at Omaezaki and Shimizu relative to Uchiura, the locations of these tide-gauge stations being shown in the insert of Fig. 3.7. As the meteorological and oceanographical disturbances to sea level are likely to be approximately eliminated by making the differences in tidal record between two stations, it is clearly seen in the figure that Omaezaki and Shimizu have been subsiding over many years. The amount of subsidence thus disclosed harmonizes with that deduced from levelling surveys.

3.1.3   North Izu area

The test survey of precise geodetic network over the North Izu area, about 100 km southwest of Tokyo, where we had the 1930 North Izu earthquake ($M = 7.0$) and the associated Tanna fault, brought out an enormous accumulation of horizontal strain in the neighborhood of the earthquake fault during 1931–73. The distributions of maximum shearing strain and principal axes for triangles over the area are shown in Fig. 3.8 (GEOGRAPHICAL SURVEY

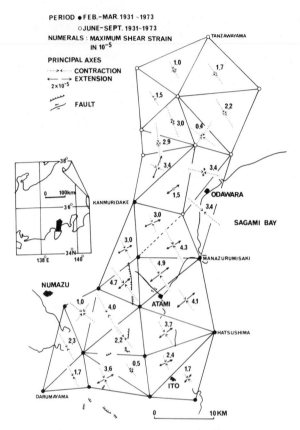

FIG. 3.8. Strain accumulation over the North Izu area during 1931–73 (GEO-GRAPHICAL SURVEY INSTITUTE, 1974a, b).

INSTITUTE, 1974a, b). Analysing the data taken by GSI, RIKITAKE (1975b) estimated that the mean accumulation rate of maximum shearing strain amounts to 0.096×10^{-5}/yr. This value is unusually high. Should the fault move again, it is anticipated that a fault movement similar to the 1930 one would occur judging from the pattern of strain distribution over the area.

3.1.4 San Andreas fault area

The only non-Japanese area over which an estimate of strain accumulation rate can be made is the San Andreas fault area, California. Fairly frequent surveys have been made over the area.

SCHOLZ and FITCH (1969, 1970) estimated the rate in an area including San Francisco, where we had an earthquake of magnitude 8.3 in 1906, as 0.05×10^{-5}/yr. The rate in the Fort Tejon area, about 100 km north of Los Angeles, was estimated as 0.06×10^{-5}/yr. An earthquake of magnitude 8 or so occurred there in 1857.

SAVAGE and BURFORD (1970, 1971) concluded, however, that the strain accumulation is not so high because of the creep movement at the fault.

3.2 Ultimate Strain of the Earth's Crust

TSUBOI (1933), who analysed then-available data of crustal movement associated with an earthquake, concluded that the earth's crust seems likely to break resulting in a large earthquake when the crustal strain exceeds an ultimate value amounting to 10^{-4} or thereabout. As the data of earthquake-associated crustal movement increased considerably in recent years, RIKITAKE (1975b) examined crustal deformation at a focal area, especially nearby an earthquake fault, for 26 inland earthquakes reaching a conclusion that the mean ultimate strain amounts to 4.7×10^{-5} with a standard deviation of 1.9×10^{-5}.

In the case of an inter-plate earthquake at a subduction zone, it is possible to estimate the ultimate displacement by multiplying the relative speed between the sea and land plates, which is obtained from other evidence, by the return period. Figure 3.9 is the histogram of such ultimate displacements as deduced from the data in the last chapter.

RIKITAKE (1976b) converted the displacement data into strain data with some plausible assumptions although the detail of such a procedure should be referred to the original paper. The ultimate strains thus estimated for earthquakes at subduction zones are shown with open columns in Fig. 3.10 in which the solid columns indicate the histogram of ultime strain for inland earthquake.

The mean ultimate strain and its standard deviation are estimated as 4.3×10^{-5} and 2.3×10^{-5}, respectively, for inter-plate great earthquakes at subduction zones. The ultimate strain and its standard deviation obtained from the whole data including both inland and

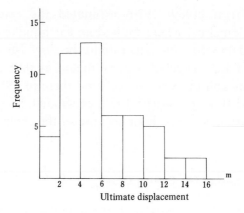

F<small>IG</small>. 3.9.   Histogram for the ultimate displacement of plate motion.

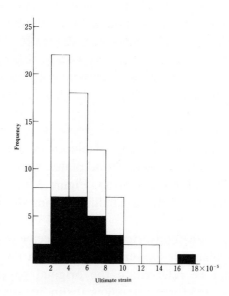

F<small>IG</small>. 3.10.   Histogram of ultimate strain for the whole data, i.e. inland and subduction-zone earthquakes. The shaded columns indicate the histogram for inland earthquake.

inter-plate earthquakes amount to 4.4×10^{-5} and 1.7×10^{-5}, respectively (R<small>IKITAKE</small>, 1976b).

3.3 Probability of Earthquake Occurrence

HAGIWARA (1974) applied a Weibull distribution analysis to the statistics of ultimate strain. Theory of Weibull distribution analysis has been presented in Subsection 2.2.3 in which time t is used as a variable. If the strain accumulation rate is denoted by u, the strain ε is given by

$$\varepsilon = ut \qquad (3.1)$$

It is therefore possible to treat the distribution of ultimate strain as shown in Fig. 3.10 as that of ultimate time. The strain accumulated at a certain epoch in a certain area can, as given in Eq. (3.1), be calculated by multiplying the strain rate u by the time span t since the last great earthquake that occurred there on the assumption that the strain energy hitherto stored was released completely at the time of that earthquake.

RIKITAKE (1975b), who determined the two parameters of a Weibull distribution representing the histogram in Fig. 3.10, estimated the cumulative probability of earthquake occurrence in the South Kanto area as shown in Fig. 3.11. In this case the time origin is taken at 1923 when we had the Kanto earthquake ($M = 7.9$). The probability amounts to only 20% at 1980, so that the author feels that the Kanto earthquake would not recur soon. However, the probability would become as high as 90% by the middle of the 21st century, so that we may well expect another large earthquake again.

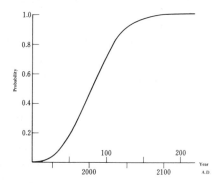

FIG. 3.11. Cumulative probability of a great earthquake recurring in the South Kanto area.

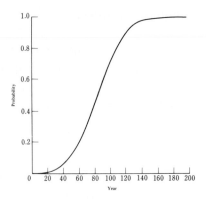

FIG. 3.12.   Cumulative probability of a great earthquake recurring in the Tokai area.

Similar estimate of probability can be made for the Tokai area by making use of the strain rate obtained in Subsection 3.1.2 as shown in Fig. 3.12. It is remarkable that the probability amounts to almost 90% at 1980 (RIKITAKE, 1977a). The time origin is taken at the year 1854 when the Ansei Tokai earthquake ($M=8.4$) occurred off the Tokai area in the Pacific Ocean.

GSI put forward a remark, however, that, if the error of triangulation survey is adjusted in such a way that the error is minimized over the Suruga Bay, the strain rate takes on a little smaller value such as 0.033×10^{-5}/yr (GEOGRAPHICAL SURVEY INSTITUTE, 1977). If we adopt this value for the strain rate, the probability becomes somewhat smaller and takes on a value around 70%. In any case the probability in the Tokai area is several times as high as that in the South Kanto area.

It is possible to estimate the probability $F_s(\varepsilon_s | \varepsilon)$ for an earthquake occurring in a strain range $\varepsilon \sim \varepsilon + \varepsilon_s$ on the condition that no earthquake occurred for $0 \sim \varepsilon$. Being called the hazard rate in the quality control technology, such a probability is given as

$$F_s(\varepsilon_s | \varepsilon) = [F(\varepsilon + \varepsilon_s) - F(\varepsilon)] / [1 - F(\varepsilon)] \tag{3.2}$$

In association with the cumulative probability in Fig. 3.12, the probability for an earthquake occurring within a 10-year period from a certain epoch can be estimated as shown in Fig. 3.13.

A similar estimate of cumulative probability of earthquake

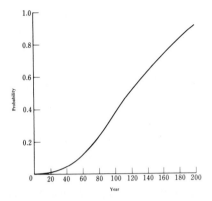

FIG. 3.13. Changes in the hazard rate, i.e. the probability of having a great earthquake in the Tokai area within 10 years from a specified epoch.

occurrence for the North Izu area indicates a high value of 85% or thereabout by the end of this century because the strain rate is anomalously high there. It has been told by KUNO (1936) that the Tanna fault, that moved by 2–3 m at the time of the 1930 North Izu earthquake, had been moving in geological time with a rate of 2 m per 1,000 years as can be inferred from a topographical observation. An earthquake having a magnitude of 7.0 or so is reported to occur in the Izu area in 841, so that the frequency of occurrence of destructive earthquakes in the area is so far believed to be once in 1,000 years. Judging from the probability discussed here, it seems likely that the hitherto-said return period of 1,000 years is too long. It should also be noted that an anomalous uplift (see Fig. 4.1) appeared towards the southern end of the Tanna fault in the middle of the 1970's.

The cumulative probabilities of an earthquake recurring in the San Francisco and Fort Tejon areas are estimated as 30 and 80%, respectively, on the basis of the strain rates obtained in Subsection 3.1.4. It is not known, however, whether the estimate of strain rate is accurate enough because of creep movement at the San Andreas fault.

It has been shown in this chapter that the probability of occurrence of a large earthquake can sometimes be estimated on the basis of the result of geodetic surveys. The point that the degree of danger can be expressed quantitatively in some cases, if not all, indicates the marked development of earthquake prediction theory in recent years. When the proposed network of precise geodetic survey is completed, a much more accurate estimate of probability of earthquake occurrence would become possible in many areas in Japan.

The estimate of probability as presented in Chapters 2 and 3 is only a kind of long-term prediction or, so to speak, regionalization. Although the probabilities estimated in these chapters are based on crustal strain accumulation and not on an actual symptom or precursor, they can be used for providing intensified measures for earthquake disaster prevention. It is reasonable to increase the budget for an area where the estimated probability of earthquake occurrence is high. Actually, the designation under the "Large-scale Earthquake Countermeasures Act" of the Tokai area to an "area under intensified measures against earthquake disaster" is undertaken on the basis of this kind of long-term prediction.

CHAPTER 4

EARTHQUAKE PRECURSORS

Many geophysical, geochemical, and biological anomalies preceding an earthquake have long been reported. As long-term prediction of earthquake occurrence has become possible, to some extent in recent years, as can be seen in Chapters 2 and 3, intensive observations of various disciplines are often being carried out over areas where the probability of an earthquake occurring in the near future is high, so that many precursory effects have come to our knowledge. The capability of acquiring data relevant to earthquake precursors have drastically increased in recent years because of the development of earthquake prediction programs mostly in Japan, the U.S.A., the U.S.S.R., and the People's Republic of China. Accordingly, the nature of precursors forerunning an earthquake is now becoming gradually more coherent.

RIKITAKE (1975c, 1976a) presented a fairly comprehensive summary of earthquake precursor data reported by 1975. In view of the rapid increase in the number of data, RIKITAKE (1979d, 1980) resummarized additional data of earthquake precursors. This chapter aims at reporting precursor data observed mainly in the 1970's. As the author attempts to avoid the overlap between the former and present data, those who are interested in relatively old data are kindly asked to refer to RIKITAKE (1975b, 1976a) or the references cited in them.

4.1   Land Deformation

It has been believed that land deformation as brought out by repeating geodetic surveys sometimes provides one of the most reliable premonitory effects of an earthquake. The most famous example is the anomalous uplift preceded the Niigata earthquake ($M = 7.5$, 1964) (DAMBARA, 1973; see RIKITAKE, 1976a, Fig. 5-6-4). The gradual change since 1900 in height at bench marks along a levelling route running along the Japan Sea coast and passing through the Niigata City changed its rate around 1954. The anomalous uplift seemed to reach its maximum around 1959, and no marked change occurred until the time of earthquake occurrence. The anomalous uplift preceded the Niigata earthquake encouraged prediction-oriented seismologists because a precursor was convincingly observed 8–9 years prior to the earthquake in this case.

During the 1965–67 swarm earthquakes at Matsushiro, Central Japan, an uplift amounting to 70 cm was observed at the center of the seismic activity (IZUTUYA, 1975) although it is not clear whether or not any precursory land deformation took place because of con-tinuous occurrences of shock. According to levelling surveys that covered a wider area around Matsushiro, however, a few uplifts observed at Toyono City and Aoki Pass did not culminate in earthquake occurrences there. TSUBOKAWA (1973) pointed out, how-ever, a magnitude 5.0 earthquake that occurred at a distance of about 40 km to the southwest of the main area of activity in 1967 was accompanied by a premonitory uplift.

In 1974, an anomalous uplift centering at Kawasaki City, a highly industrial area adjacent to Tokyo, was detected (GEOGRAPHICAL SURVEY INSTITUTE, 1975). Although no official com-ment on possible occurrence of an earthquake was made, social unrest has arisen among local people who were afraid of earthquake hazards in such a densely-populated area in case of earthquake occurrence there. But, no earthquake has so far occurred. Although the anom-alous uplift was likely to be caused by something anomalous related to underground water, nothing is known about the real cause of the uplift.

According to SATO and INOUCHI (1977), examples of anomalous

uplift that precedes an earthquake are not many. Only 17% of uplifts is connected to an earthquake occurrence some time later. It is not always correct, therefore, to assume an earthquake occurrence whenever an uplift of ground is observed.

The author should here like to point out the fact that there are two well-known uplifts at the moment. As is mentioned in Subsection 1.2.2 and shown in Fig. 1.19, a widespread uplift has been found in southern California originating and centering at Palmdale. The now-famous Palmdale Bulge may or may not be connected to earthquake occurrence. But a number of U.S. seismologists seem to be inclined not to believe the occurrence of large earthquake. It has even been argued that the apparent uplift is caused by the accumulation of measuring error as mentioned in Subsection 1.2.2.

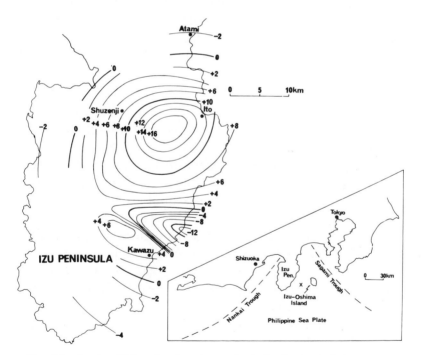

FIG. 4.1. Land uplift in units of centimeters in the Izu Peninsula during a period from 1967–69 to 1978 (GEOGRAPHICAL SURVEY INSTITUTE, 1978). The 1978 survey was carried out immediately after the Izu-Oshima Kinkai earthquake of magnitude 7.0 that occurred on Jan. 14, 1978. The geography around the peninsula is shown in the insert in which the epicenter of the above earthquake is shown with a cross.

In contrast to such a large-scale uplift in California, a more localized uplift in the Izu Peninsula, about 100 km southwest of Tokyo, has been noticed since around 1975. Figure 4.1 shows the uplift during a period from 1976–69 to 1978 (GEOGRAPHICAL SURVEY INSTITUTE, 1978). The last survey being carried out immediately after the January 14, 1978 Izu-Oshima Kinkai earthquake ($M = 7.0$), of which the epicenter is indicated with a cross in the insert of Fig. 4.1, some subsidence associated with the fault appeared at the time of main shock can be seen on the east coast of the peninsula.

However, the uplift, that had been taking place towards the west of Ito City, was very little affected by the earthquake. It seems likely that the uplift developed mostly at around 1974–75. A magnitude 6.9 earthquake occurred at the extremity of the peninsula in May, 1974. Since then, the peninsula became tumultuous in the geoscientific sense. During 1975–76, a series of swarm earthquakes occurred to the south of the anomalous uplift including a magnitude 5.4 shock near Kawazu. This activity was followed by the Izu-Oshima Kinkai earthquake in January, 1978.

Towards the end of the year 1978, a magnitude 5.0 earthquake occurred again near Kawazu, and swarm earthquakes occurred a few kilometers off Kawanazaki near Ito including a magnitude 5.4 shock in December. The GEOGRAPHICAL SURVEY INSTITUTE (1979b), hurriedly relevelled the route from Atami to Kawazu via Ito. The difference between the two surveys with a time-interval of only 6 months indicates very clearly an upheaval of ground at a portion of the route close to the seismic area as can be seen in Fig. 4.2.

Looking at Fig. 4.2, we feel that the pattern of uplift astonishingly resembles that associated with the Ito earthquake swarm in 1930 (TSUBOI, 1933; RIKITAKE, 1976a, Fig. 5.48). As the North Izu earthquake ($M = 7.0$, November 26, 1930) occurred about 6 months later the swarm activity, CCEP, which also pays attention to the strain accumulation in the North Izu area as shown in Fig. 3.8 and the existing uplift to the east of Ito as shown in Fig. 4.1, is intensifying observations of various disciplines over the area concerned. Swarm activities recurred at almost the same area in March and May, 1979 although the intensity of activity is somewhat weaker than that for the swarm last year.

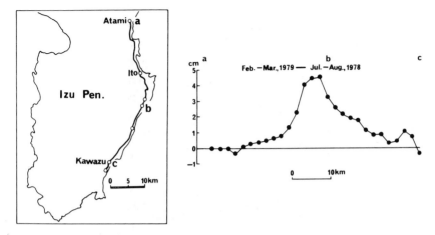

F<small>IG</small>. 4.2.   Ground uplift during a period from July–Aug., 1978 to Feb–Mar., 1979 as revealed by levelling surveys along the route connecting Atami to Kawazu via Ito (G<small>EOGRAPHICAL</small> S<small>URVEY</small> I<small>NSTITUTE</small>, 1979b).

In June 1980, the swarm activity came back underneath the sea off Kawanazaki including a magnitude 6.7 shock. The activity was followed by a swarm of small earthquakes occurring off Manazuruzaki some 20 km north of the epicenters of the last activity. Although the northward shift of seismic activity in the Izu area is apparent, its physical meaning is not entirely clear.

4.2   Change in Sea Level

A number of sea retreats forerunning an earthquake along with premonitory changes in sea level observed by a tide-gauge have been reported. Those relatively old data are summarized by R<small>IKITAKE</small> (1975c, 1976a).

Additional evidence that the sea level indicated a conspicuous change 1.5 days prior to the 1946 Nankai earthquake ($M = 8.1$) is put forward by S<small>ATO</small> (1977) who recently analysed hourly data of tide gauges at Hosojima and Tosashimizu respectively located on the coast of Kyushu and Shikoku, Japan. As can be seen in Fig. 4.3, the sea level at the latter station about 230 km distant from the epicenter

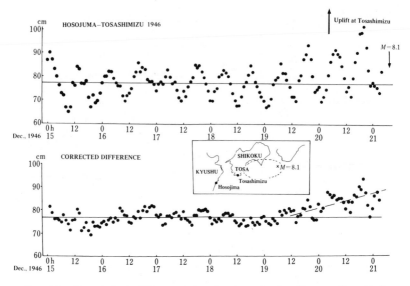

FIG. 4.3.   Changes in the difference in hourly values of sea level between Tosashimizu and Hosojima before the Nankai earthquake ($M=8.1$, 1946). The lower graph is corrected for the effect of tidal current (SATO, 1977).

seems to have recorded an anomalous land upheaval amounting to 50 cm in total.

WYSS (1975, 1976a, b, 1977) examined changes in sea level concluded that precursory effects were found for the Antofagasta, Chile, earthquake ($M=6.8$, 1960), Peru earthquake ($M=7.5$, 1966), Hyuganada, Japan, earthquake ($M=7.5$, 1968), and Long Beach, California, earthquake 2,190, 1,830, 1,280, and 1,800, days preceding the main shock, respectively. The changes in sea level ranged 3 to 8 cm.

4.3   Tilt, Strain, and Crustal Stress

As numerous examples of precursory tilt and strain are reviewed in RIKITAKE (1975c, 1976a), only additional data that came to the author's knowledge quite recently will be mentioned in this section.

In China, levelling surveys over a distance of several hundreds or tens of meters are popular. These short-distance levellings are useful for detecting a local change in ground tilting. A pair of short-distance levelling routes has been set up in a L-shape at Jin Xian about 185 km southwest of the epicenter of the Haicheng earthquake ($M = 7.3$, 1975). The route that crosses the Jin Zhou fault striking N 15°E has a length of 560 m, while another route parallel to the fault has a length of 360 m (RALEIGH et al., 1977). The end bench marks of these base-lines are concrete-blocks 2 m in length buried in the ground. While, those bench marks between the end marks consist of concrete-blocks 1 m in length. The ground covered by the routes is almost flat with undulations of a few meters.

Levelling surveys of the first order have been carried out in regard to the short-distance levelling routes every day since 1971. Although some seasonal changes have been observed, they are eliminated from the graph which shows the change in ground tilting. Changes in the difference in height between end points B and A and B and C are shown in Fig. 4.4 in which the locations of these end points are also shown.

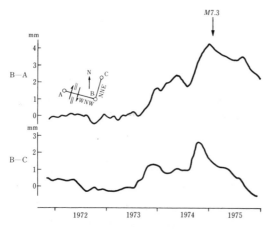

FIG. 4.4. The changes in relative elevation as a function of time for 1972–76 along a short levelling route across the Jin Zhou fault at Jin Xian (RALEIGH et al., 1977).

It is remarkable that a tilting towards a northwest direction with a rate of 10 μrad/yr started in the later half of 1973. The data at intermediate bench marks between the end points indicate that the tilting is almost uniform and that no discontinuous movement is observed across the fault. It is said that the change in tilt as can be seen in Fig. 4.4 plays an important role on the prediction of the

Haicheng earthquake ($M = 7.3$, February 4, 1975). Especially, the reversal of tilting direction towards the late 1974 was important for the short-term prediction.

At the Shenyang Seismological Observatory, about 100 km north-northeast of the epicenter, a reversal of ground tilting was also noticed in December, 1974.

An indoor base line for extremely short-distance levelling has been set up at the Dahuichang Crustal Movement Observatory about 25 km southwest of Peking (Beijing). The base line crosses the Papaoshan fault, one of the main active faults in North China although the length of the line amounts to only several tens of meters. As can be seen in Fig. 4.5, an observer reads the scale of a mark, which is shown in Fig. 4.6 with a Zeiss Ni004 level. He then makes a 180 degree turn of the level and reads the scale of another mark. Making the difference in scale reading between the two marks, it is possible to observe the ground tilting. Measurements are usually made every 4 hours.

A water-tube tiltmeter and a strainmeter are set up in the basement of the observatory as can be seen in Fig. 4.7. Figure 4.8 shows the changes in ground tilting as observed with the water-tube tiltmeter and short-distance levelling at the Dahuichang Observatory along with the short-distance levelling at Niuk'ouyu about 20 km further southwest from Dahuichang. The changes in Fig. 4.8 may possibly be classified into the medium- and short-term precursors to the Tangshan earthquake ($M = 7.8$, 1976) respectively starting from the middle of the year 1975 and around June, 1976. The curves in Fig. 4.8 exhibit a similarity to those in Fig. 4.4 which shows the premonitory tilt change at the time of the Haicheng earthquake.

Many tiltmeters are in operation in California as can be seen in Figs. 1.20 and 1.21. According to JOHNSTON (1978b) a remarkable change in tilt occurred in association with a magnitude 4.3 earthquake that occurred near the Briones Reservoir on January 8, 1977. The epicentral distance was 5.5 km. The change, which is shown in Fig. 4.9, is characterized by a large deviation amounting to 2 μrad from the general change. Such a large change has not been observed during the last 3 years. It seems highly likely that the change is a kind of earthquake precursor.

FIG. 4.5.   The Zeiss Ni004 level for the indoor short levelling at the Dahuichang
Crustal Movement Observatory.

FIG. 4.6.   The scale for the indoor short levelling at the Dahuichang Crustal
Movement Observatory.

FIG. 4.7.   The water-tube tiltmeter and strainmeter at the Dahuichang Crustal
Movement Observatory.

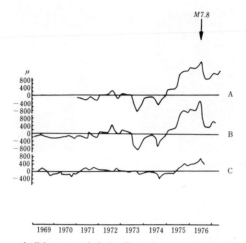

FIG. 4.8. Ground tilting preceded the Tangshan earthquake ($M = 7.8$, 1976). A: Changes in ground tilt as observed at the Dahuichang Crustal Movement Observatory as measured by the water-tube tiltmeter. B: The same observed by the short levelling. C: The result of short levelling at Niuk'ouyu. (After the Dahuichang Crustal Movement Observatory).

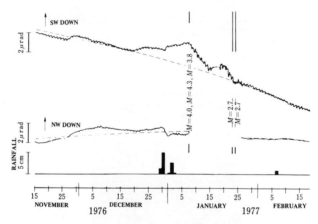

FIG. 4.9. The precursory change in ground tilting preceded the Briones, California, earthquake ($M = 4.3$, 1977) (JOHNSTON, 1978b).

In Subsection 1.1.2(4), it is stated that JMA has set up bore-hole volume strainmeters at 31 stations as shown in Fig. 1.9.

Figure 4.10 shows the precursor-like change recorded by a volume strainmeter buried at Irozaki, the southernmost tip of the Izu

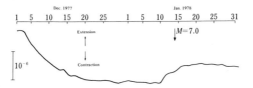

FIG. 4.10.   Changes in the record of a volume strainmeter at Irozaki at the tip of the Izu Peninsula (JAPAN METEOROLOGICAL AGENCY, 1978).

Peninsula. About 42 days prior to the Izu-Oshima Kinkai earthquake ($M = 7.0$, 1978), the strainmeter started to record a contraction as can be clearly seen in Fig. 4.10. Such a change tended to alter its sign towards expansion about 4 days prior to the main shock. At about the same period, swarm activity of small earthquakes becomes conspicuous underneath the sea bottom between Izu-Oshima Island and Izu Peninsula. The distance between the epicenter of the main shock and Irozaki amounts to 32 km.

As no protracted observations with the JMA-type volume strainmeter has been experienced, it is hard to evaluate the change as shown in Fig. 4.10. However, it has been reported that a few precursory changes have been observed by strainmeters of the same kind in association with much smaller earthquakes, so that it is not unreasonable to think that the strainmeter may be useful for detecting earthquake precursors.

In situ measurements of crustal stress are being made in China (TANAKA, 1978a, b). Those who are interested in details of measuring method, are asked to refer to TANAKA (1978a, b). It has been reported that the crustal stress at a point 50 km distant from the epicenter of the Haicheng earthquake ($M = 7.3$, 1975) tended to increase 7 months prior to the earthquake reaching the maximum value amounting to 4.31 kg/cm^2. The direction of principal axis changed suddenly about 1 month preceding the main shock.

In situ measurement of crustal stress has now become widely made in the U.S.A. (e.g. McGARR and GAY, 1978; JAMISON and COOK, 1980) and Japan (e.g. NATIONAL RESEARCH CENTER FOR DISASTER PREVENTION, 1979; RESEARCH GROUP FOR CRUSTAL STRESS IN WESTERN JAPAN, 1980; GEOLOGICAL SURVEY OF JAPAN, 1980, 1981).

4.4 Seismological Precursor

4.4.1 Foreshock

Figure 4.11 (ZHU, 1976) shows the hourly frequency of fore-shocks before the Haicheng earthquake ($M = 7.3$, February 4, 1975). Small earthquakes had begun to be felt from February 1. The number and magnitude increased day by day. An enormous increase in the hourly frequency was observed from the afternoon of February 3. But the number of shocks suddenly tended to decrease in the morning of February 4. Magnitude 4.7 and 4.2 shocks occurred on that morning. The main shock occurred at 19h 36m in the evening. Such a foreshock activity played an important role on the imminent prediction.

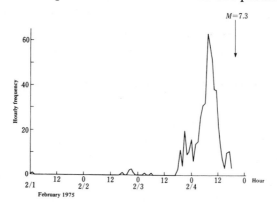

FIG. 4.11.   Hourly frequency of foreshocks to the Haicheng earthquake (ZHU, 1976).

In relation to the Izu-Oshima Kinkai earthquake ($M = 7.0$) that occurred on 12h 24m on January 14, 1978, the following seismic activity has been reported. According to JMA's observation, commencement of a swarm activity of small earthquakes around Izu-Oshima Island had been recognized from January 12. The area in-between Izu-Oshima Island and Izu Peninsula is noted for occasional occurrences of earthquake swarms although the activity appeared to be a little high this time. As can be seen in Fig. 4.12 (JAPAN METEOROLOGICAL AGENCY, 1978), the number of small shocks began to increase so enormously that we observed almost continuous vibrations on the seismogram from 09h 30m in the morning of January

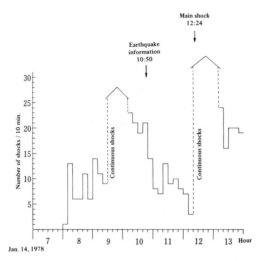

FIG. 4.12. Frequency per 10 min of foreshocks to the Izu-Oshima Kinkai earthquake as observed on Izu-Oshima Island (JAPAN METEOROLOGICAL AGENCY, 1978).

14. However, a drastic decrease in the number of shocks began to be noticed at around 10 o'clock.

In light of such development of swarm activity, JMA issued an earthquake information to the public at 10h 50m. The information reads "A moderately large earthquake, that gives rise to slight damage, may occur shortly" (see Table 8.2). About 1.5 hours after the information issuance, the main shock actually occurred somewhere between Izu-Oshima Island and Izu Peninsula at 12h 24m.

Both the cases of Haicheng and Izu-Oshima Kinkai earthquakes provide typical examples of a characteristic pattern of foreshock sequences, namely an enormous increase in earthquake number followed by a sudden decrease culminating in the occurrence of main shock.

4.4.2 Anomalous seismic activity

SEKIYA (1976, 1977) pointed out a tendency that a swarm-like seismic activity sometimes takes place in an area in which a large earthquake occurs later on. Figure 4.13 shows the cumulative frequency of earthquake occurrence in the epicentral area of the Gifu Chubu earthquake ($M=6.6$, 1969). Very low seismicity had been

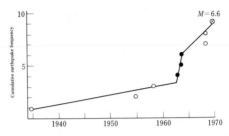

FIG. 4.13. Anomalous seismic activity preceded the Gifu Chubu earthquake ($M = 6.6$, 1969) (SEKIYA, 1976).

noticed in the area concerned over a long period of time. However, three earthquakes occurred successively in a 1.3-year period after a magnitude 4.3 earthquake that occurred on April 1, 1962. The main shock occurred in the area about 6.5 years later. Such anomalous seismic activities can be observed in the case of Fukui ($M = 7.3$, 1948), North Miyagi ($M = 6.5$, 1962) and Shizuoka ($M = 6.1$, 1965) earthquakes. SEKIYA (1976) could identify 10 earthquakes accompanied by such preceding seismic activity.

EVISON (1977a, b) also pointed out similar precursory seismicity for 4 and 9 earthquakes respectively in California and New Zealand. The logarithm of precursor time T of these anomalous activity seems likely to be proportional to the magnitude M of main shock. The relation between $\log_{10} T$ and M roughly agrees with that in RIKITAKE (1975c, 1976a, 1979f). This seems to have an important significance for earthquake prediction.

ISHIDA and KANAMORI (1977) studied an anomalous seismic activity preceding the San Fernando earthquake ($M = 6.4$, 1971). MARZA(1979) pointed out that even an intermediate-depth earthquake, the Vrancea, Romania, earthquake ($M = 7.2$, 1977), was accompanied by an anomalous activity. The precursor time agrees well with Rikitake's relationship.

4.4.3   Seismic gap of the second kind

The concept of seismic gap has already been introduced in Section 2.3. An area in which the activity of small earthquakes has decreased prior to occurrence of a large earthquake is named a seismic gap of the second kind (MOGI, 1978).

OHTAKE (1976) and SEKIYA (1976) noticed that a fairly wide area including Izu Peninsula was seismically calm during the period of several years before the Izu-Hanto Oki earthquake ($M=6.9$, 1974).

OHTAKE et al. (1977) pointed out that two Oaxaca, Mexico, earthquakes of magnitude 7.3 (1965) and 7.1 (1968) were accompanied by a seismic gap of the second kind. The characteristics of these gaps are as follows;

1)   A clear-cut period of gap followed by a recovery period of seismicity was observed.

2)   The precursor time from the beginning of the gap to the occurrence of main shock was about 2 years.

The epicenters of main shocks were located at the northern-most periphery of the aftershock areas. It seems likely that the rupture, that gave rise to the main shock, began at the land-side extending towards the sea-side. The main shocks were caused by low-angle thrust faults that are typical at a subduction zone.

The gap, that preceded the August 2, 1968 earthquake ($M=7.1$), is shown in Fig. 4.14. It is seen in the figure that the epicentral area has been quiet during a period of about 2 years.

OHTAKE et al. (1977) further pointed out that an area between the focal areas of the 1965 and 1968 earthquakes had been in a state of conspicuous gap since June, 1973. In Fig. 4.15, they showed that a well-recognized gap can be observed during a period from June, 1973 to May, 1975.

On the basis of the above facts, OHTAKE et al. (1977) pointed out

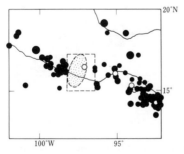

FIG. 4.14.   The seismic gap near Oaxaca, Mexico precursory to the Aug. 2, 1968 earthquake ($M=7.1$). The open circles indicate the earthquakes shallower than 60 km after the recovery of seismic activity. The ellipse shows the aftershock area of the main shock (OHTAKE et al., 1977).

a possibility of a large earthquake occurring in the Oaxaca area. The magnitude is supposed to take on a value of 7.5 with an allowance of ± 0.25, and the supposed location of epicenter is (16.5°N, 96.5°W) with a probable error of $\pm 0.5°$ for latitude and longitude. There was no guess about the time of earthquake occurrence although it was mentioned in the paper that the gap had been persisted during a period which is longer than that for the past earthquakes in the Oaxaca area.

The above work by OHTAKE et al. (1977) was distorted by non-scientific predictors who sent the Mexican President a message that foretold the occurrence of a large earthquake at Pinotepa in Oaxaca Province on April 23, 1978. As the foretelling leaked out, a panic started among the local inhabitants resulting in voluntary evacuation and other crazy unrest. Even a false rumor that nuclear bombs were set at undersea faults was circulated.

GARZA and LOMNITZ (1978), who assumed that earthquakes are occurring randomly and stationarily, checked the significance of seismic gap reported by OHTAKE et al. (1977), and reached a conclusion that such a gap is not significant from the viewpoint of statistics based on the Poisson distribution. It appears to the author that recurrence of a large earthquake at such a subduction zone is readily expected as we have seen in Chapter 2. The assumption that Poisson's law holds good for earthquake occurrence may not be applicable to such a particular earthquake area in a subduction zone.

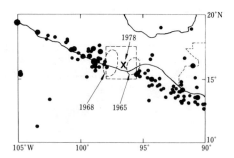

FIG. 4.15.   The seismic gap near Oaxaca, Mexico during June, 1973–May, 1975. Solid circles indicate earthquakes shallower than 60 km. The two ellipses are the aftershock areas of the 1965 ($M = 7.3$) and 1968 ($M = 7.1$) earthquakes. The corss shows the epicenter of the Nov. 29, 1978 earthquake as inserted by the present author (OHTAKE et al., 1977).

According to NOAA, an earthquake of magnitude 7.8 (probably surface wave magnitude) occurred at (16.07°N, 96.49°W) at 19h 52m 53.2s (UT), November 29, 1978 as predicted. The magnitude and location of epicenter agree very well with those guessed by OHTAKE *et al.* (1977). The epicenter is shown with a cross in Fig. 4.15. The precursor time is estimated as 5.4 years in this case.

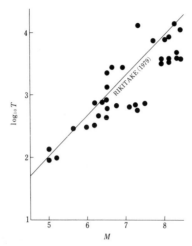

FIG. 4.16. The relation between the logarithmic precursor time *T* measured in units of days and the magnitude *M* of the main shock. The straight line is due to Rikitake's empirical relation (RIKITAKE, 1979d). It is inserted by the present author in place of the original line based on RIKITAKE (1975c, 1976a). (OHTAKE *et al.*, 1978).

OHTAKE *et al.* (1978) examined the duration time of seismic gap of the second kind for earthquakes in Mexico, Central America and California. The relation between duration time and magnitude of main shock obtained by them is shown in Fig. 4.16. WYSS *et al.* (1978) conducted a similar study.

4.4.4 Growth and decay of seismic activity

What are mentioned in Subsections 4.4.1, 4.4.2, and 4.4.3 are some of distinct features of precursory seismic activity. In this subsection, growth and decay of seismic activity as a precursory indicator will be discussed in a more general way.

RIKITAKE (1976a) presented a number of premonitory increases in seismicity in a fairly wide area including the epicenter of main

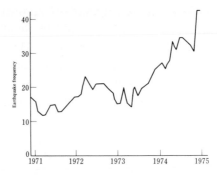

FIG. 4.17.   Monthly number of earthquakes in southern Liaoning Province (ZHU, 1976).

shock. In Fig. 4.17 (ZHU, 1976), the premonitory increase in the number of South Liaoning Province earthquakes that appears to begin about 1.5 years prior to the Haicheng earthquake ($M = 7.3$, 1975) is shown. In general, there is a tendency that the seismicity first increases in a wide zone outside the epicentral area, and that a calm period similar to seismic gap of the second kind persists for a certain period of time. When earthquakes start to occur in that gap, it is anticipated that a large earthquake will shortly occur. However, it cannot be said that the relationship between such seismic activity and occurrence of a large earthquake is well established. There still remain many things to be studied in relation to the subject.

As has been mentioned in association with the Haicheng ($M = 7.3$, 1975) and the Izu-Oshima Kinkai ($M = 7.0$, 1978) earthquakes, it has been noticed that premonitory seismicity is suddenly quieted a few hours prior to the main shock. Such a phenomenon might be accounted for by dilatancy hardening (see Section 4.12, SCHOLZ et al., 1973) although no definite proof has been made.

4.4.5   b value

The number of occurrense (N) for earthquakes of magnitude M in a region during a certain period is correlated to M as

$$\log_{10} N = a - bM \qquad (4.1)$$

where a and b are constants. Equation (4.1) is called the Gutenberg-Richter formula (GUTENBERG and RICHTER, 1944).

In RIKITAKE (1976a) are quoted many an example of decrease in *b* value forerunning an earthquake. HASEGAWA *et al.* (1975) reported on a decrease in *b* value from 1.0 to 0.8 about 730 days prior to the Southeast Akita Prefecture, Japan, earthquake (*M* = 6.2, 1970). It is also reported by the NATIONAL RESEARCH CENTER FOR DISASTER PREVENTION (1978) that *m* value (*m* = *b* + 1, ISHIMOTO and IIDA, 1939) for the foreshocks of the Izu-Oshima Kinkai earthquake (*M* = 7.0, 1978) is anomalously small amounting to only 1.551 or thereabout.

Many attempts to make use of *b* value for earthquake prediction have recently been made in China. In relation to the 1976 Longling earthquakes (*M* = 7.5, 7.6), TANG (1978) presented the secular change in *b* value in the western part of China as shown in Fig. 4.18. Attention should be drawn to the decrease in *b* value since 1965. LI *et al.* (1978) and MA (1978) examined changes in *b* value associated with large earthquakes in China. Figure 4.19 shows the change in *b* value in the Tangshan area (LI *et al.*, 1978). The decrease preceding the Tangshan earthquake (*M* = 7.8, 1976) is outstanding. In China, it is understood that *b* value is an indicator of the general stress state in a focal region. The decrease in *b* value indicates that the stress approaches the ultimate value.

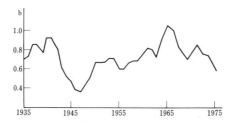

FIG. 4.18.   Changes in *b* value in western China (TANG, 1978).

FIG. 4.19.   Changes in *b* value in the Tangshan area before the Tangshan earthquake (*M* = 7.8, 1976), the occurrence time of which is shown with an arrow (LI *et al.*, 1978).

4.4.6 Source mechanism

Much of precursory change in source mechanism as found in the Garm area of the U.S.S.R. was summarized in RIKITAKE (1976a). GUPTA (1975) reported on a reversal of initial motion of P wave 38 days prior to a magnitude 4.0 earthquake near Slate Mountain, Nevada, the U.S.A. Some more work related to precursory change in source mechanism have been reported in recent years. They are LINDH et al. (1978) on precursory change in source mechanism and ISHIDA and KANAMORI (1978) on change in wave form.

4.4.7 Hypocentral migration of microearthquakes

GUPTA (1975) noticed that the hypocenters of microearthquakes moved 4–6 km preceding the Nevada earthquake mentioned in the last subsection. That the origins of microearthquakes became deep by 25% 60 days preceding a magnitude 4.6 earthquake at Stone Canyon, Central California, was reported by BUFE et al. (1974). According to ENGDAHL and KISSLINGER (1977), a several kilometer migration of microearthquake hypocenters toward the focus of a magnitude 5.0 earthquake was observed in Mid Aleutian Islands 35 days prior to the occurrence of main shock.

4.4.8 Change in seismic wave velocity

Changes in seismic wave velocities, especially in the ratio of P wave velocity (V_P) to S wave velocity (V_S) have been thought to provide a powerful means of earthquake predition since the U.S.S.R. finding (SEMYENOV, 1969) in Garm and the following work. As it appeared that premonitory change in V_P/V_S is a logical conclusion of a dilatancy model (SCHOLZ et al., 1973), change in seismic wave velocities drew much attention of prediction-oriented seismologists.

Accordingly, many tests for checking the validity of premonitory change in V_P, V_S, or V_P/V_S ratio have been made by many seismologists (TERASHIMA, 1974; UTSU, 1975; OHTAKE, 1976; IIZUKA, 1976a, b, c; WANG et al., 1976; FENG et al., 1974, 1976a, b; FENG, 1975; DUAN et al., 1976). RIKITAKE (1979d) summarized 17 examples of changes in V_P/V_S ratio reported from China. Figure 4.20 is the change in V_P/V_S ratio forerunning the Longling earthquakes ($M = 7.5$, 7.6, 1976) (TANG, 1978). Most Chinese reports lack estimates of accuracy.

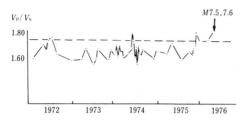

FIG. 4.20. Changes in V_P/V_S forerunning the Longling earthquake ($M=7.5$, 7.6, 1976) (TANG, 1978).

OHTAKE and KATSUMATA (1977) and YOSHII (1978) pointed out that the accuracy of seismic observation required for detecting changes in seismic wave velocities should be unexpectedly so high that many of the results reported thus far are not reliable. It turns out that it is no easy matter to detect changes in seismic wave velocities in spite of the optimistic view which was popular several years ago. There are many cases for which we have no changes in seismic wave velocities in the western United States as reported by CHOU and CROSSON (1978).

WYSS (1981) believes that the changes associated with the events in the Blue Mountain Lake area (AGGARWAL et al., 1975) and Hawaii (JOHNSTON, A.C., 1978) are reliable. He also pointed out that no precursory change in seismic wave velocity is observed for earthquakes caused by strike-slip and normal faults. OHTAKE and KATSUMATA (1977) pointed out the change forerunning the North Mino earthquake ($M=7.0$, 1961) is reliable.

It has been known from rock breaking tests that dilatancy is produced immediately prior to rupture resulting in a decrease in P wave velocity (e.g. MATSUSHIMA, 1960). If so, it is rather puzzling that no change in seismic wave velocity is associated with many natural earthquakes. We have to apply the results of laboratory experiment to natural phenomena with much care.

4.5 Earth Tide

In addition to the literature quoted by RIKITAKE (1976a), two more reports on precursory changes in earth tidal amplitude are now available. LATYNINA and RIZAEVA (1976) reported that the amplitude of M_2 constituent decreased by 6% 30 days prior to a magnitude 4.5 earthquake that occurred nearby Dushambe in 1976. According to MIKUMO *et al.* (1977), who analysed the data observed at the Kamitakara Crustal Movement Observatory 60 km distant from the epicenter of the Gifu Chubu earthquake ($M = 6.6$, 1969), the earth tidal admittance increased by 15% 11 months preceding the earthquake.

4.6   Geomagnetic and Geoelectric Precursor

4.6.1   *Absolute value of the geomagnetic field*

SMITH and JOHNSTON (1976) reported on a remarkable change in the geomagnetic total intensity in association with a magnitude 5.2 earthquake on November 28, 1974 that occurred near Hollister, California. As can be seen in Fig. 4.21, a distinct change of 1γ or a little larger was observed about 2 months before the shock. The anomalous change in the geomagnetic field was followed by a fault creep and a change in ground tilting. As the standard deviation of magnetic noise around this area is only 0.5γ or so, the change observed must be significant and has something to do with earthquake occurrence or fault creep.

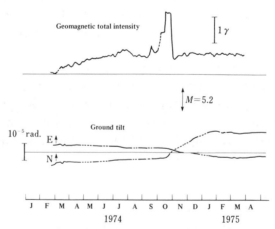

FIG. 4.21.   Changes in the geomagnetic total intensity and ground tilting preceding the Nov. 28, 1974, earthquake of magnitude 5.2 that occurred near Hollister, California (SMITH and JOHNSTON, 1976).

The recent crustal movement and seismic activity in the Izu Peninsula about 100 km or a little more southwest of Tokyo has been described in fair detail in Section 4.1 in relation to the anomalous uplift in the middle of the peninsula. Geomagnetic work aimed at monitoring possible seismomagnetic effect and geomagnetic precursor, if any, has been conducted very intensively over the peninsula. The GEOMAGNETIC SURVEY PARTY (1977) and SASAI and ISHIKAWA (1978) reported on fluctuations of total intensity at Sugehiki station (see Fig. 4.23) during 1976–77. It appears that an increase in the total intensity of a few gammas took place about one month preceding a swarm activity of earthquakes including a few shocks having a magnitude of 5 or thereabout.

The seismic activity in the Izu Peninsula culminated in the Izu-Oshima Kinkai earthquake ($M = 7.0$, 1978). HONKURA (1978a) pointed out that a precursor-like change of several gammas can be seen in the upper graph in Fig. 4.22 which shows the difference in the total intensity between Sugehiki and Matsuzaki (see Fig. 4.23). The precursor time amounts to 64 days. The seismic activity in the Izu Peninsula did not cease in spite of the Izu-Oshima Kinkai earthquake. A precursor was clearly observed before the November, 1978, Kawazu earthquake of magnitude 5.0 (GEOMAGNETIC SURVEY PARTY, 1979). The observation had been conducted at a station located almost at the epicenter. Judging from the differences in the total

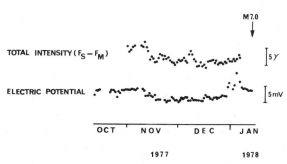

FIG. 4.22. Changes that seem likely to precede the Izu-Oshima Kinkai earthquake ($M = 7.0$, 1978). The upper graph shows the changes in the difference in the geomagnetic total intensity between F_S (Sugehiki) and F_M (Matsuzaki). The locations of the stations are shown in Fig. 4.23. The graph at the bottom shows the changes in earth potential difference as observed at Nakaizu which is close to Sugehiki (HONKURA, 1978a).

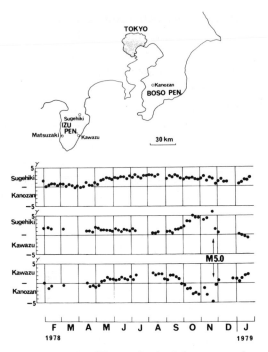

Fig. 4.23. Changes in the difference in the 5-day means of geomagnetic total intensity between the three stations; Kawazu, Sugehiki and Kanozan (Geomagnetic Survey Party, 1979).

intensity between respective stations as shown in Fig. 4.23, it is obvious that the magnetic field at Kawazu was subjected to a premonitory decrease in the total intensity starting two months before the shock.

Apart from changes in the geomagnetic field that are associated with individual earthquakes, fluctuations of the total intensity in the Izu Peninsula seems to reflect the growth and decay of the stress that is given rise by sporadic propelling of the Izu Peninsula due to the north-westward motion of the Philippine Sea plate.

Geomagnetic observations aiming at detecting precursory changes are popular in China. It was reported that the geomagnetic field at Dalian, Liaoning Province, had appreciably changed relative to that at Peking (Beijing). Comparing the vertical intensity at Dalian on May 22, 1974 to that on October 27, 1973, an increase amounting

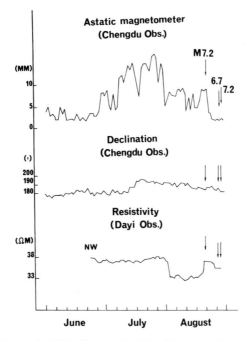

Fig. 4.24.   Top and middle: The record obtained by an astatic magnetometer and a declination variometer at the Chengdu Standard Observatory. Bottom: Changes in resistivity at the Dayi Observatory.

to 21.5 γ was observed (ZHU, 1976; RALEIGH et al., 1977).

As a result of comparing the geomagnetic field at 19 magnetic stations in September, 1974 to that in February–March, 1975, it was known that a decrease amounting to 7 to 9 γ had been taking place at stations around Tangshan. This might be a precursor of the Tangshan earthquake ($M = 7.8$, 1976) (RALEIGH et al., 1977).

At the Chengdu, Sichuan Province, Standard Seismological Observatory, Chinese colleagues are carrying out geomagnetic and geoelectric observations of various types. Figure 4.24 shows the changes recorded by an astatic magnetometer and a declination variometer along with the changes in ground resistivity observed at Dayi about 50 km west of Chengdu. These changes are believed to be forerunners of the Songpan-Pingwu earthquakes ($M = 7.2$, 6.7, 7.2, 1976) that occurred in the northern Sichuan Province at an epicentral distance of 220 km from Chengdu.

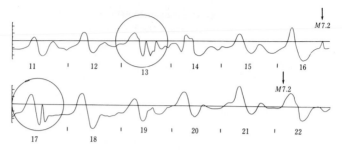

FIG. 4.25. Changes in geomagnetic declination observed at the Min Jiang gear factory during Aug. 11–22, 1976.

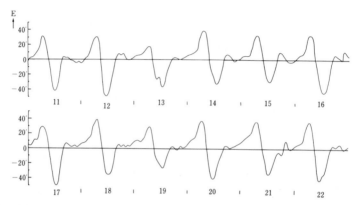

FIG. 4.26. Changes in geomagnetic declination observed at the Kakioka Magnetic Observatory during Aug. 11–22, 1976.

A simple declination variometer (see Fig. 1.37) at an amateur station at the Min Jiang gear factory in Peng Xian, Sichuan Province recorded changes that are shown in Fig. 4.25. The units of ordinate is not available. In the figure we can see two anomalous changes of spiky shape on August 13 and 17 superposing on the regular daily variation. Figure 4.26 shows the changes in declination for the same period at the Kakioka Magnetic Observatory, Japan. As no spiky changes are found on the Kakioka magnetogram, it is certain that some anomalous changes took place at the Min Jiang station although we have no guarantee that those changes are precursors.

WYSS (1975) found that the horizontal intensity of geomagnetic field decreased by $20\,\gamma$ or so 9.5 and 7.5 years preceding the Yakutat,

Alaska ($M = 7.9$, 1958) and Sitka, Alaska ($M = 7.2$, 1972) earthquakes, respectively.

4.6.2 Amplitude of short-period geomagnetic variation

According to Honkura (personal communication, 1977), the amplitude of short-period geomagnetic variations at a station close to the epicenter of the Southeast Akita Prefecture earthquake ($M = 6.2$, 1979) changed to some extent relative to the variations at a standard observatory several tens of kilometers distant from the station forerunning the earthquake. Such a change can be accounted for provided the electric conductivity in the vicinity of the focal region becomes high and so the electric currents induced in the earth by geomagnetic variations are deflected.

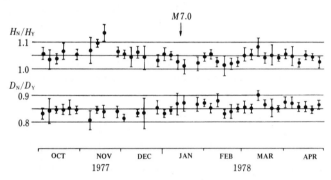

FIG. 4.27. Changes in the amplitude of short-period geomagnetic variations that seem likely to precede the Izu-Oshima Kinkai earthquake ($M = 7.0$, 1978). H_N and D_N are the amplitudes of geomagnetic variation of the horizontal intensity and declination observed at Nakaizu. H_Y and D_Y are those at the Yatsugatake Magnetic Observatory (HONKURA and KOYAMA, 1978).

HONKURA and KOYAMA (1978), who made an observation at a station close to Sugehiki (see Fig. 4.23) found a precursory change in the amplitude of short-period geomagnetic variation as can be seen in Fig. 4.27. The increase in the amplitude relative to the variations at the Yatsugatake Magnetic Observatory, about 130 km distant from the station, seems likely to have started some 70 days preceding the Izu-Oshima Kinkai earthquake ($M = 7.0$, 1978). Judging from the standard deviations also indicated in Fig. 4.27, the increase in the amplitude some time in November, 1977 may be significant. A sharp decrease in the amplitude occurred after the earthquake.

4.6.3 CA transfer function

It is known that the following relation approximately holds good between short-period variations of horizontal intensity (ΔH), declination (ΔD), and vertical intensity (ΔZ) at a certain station:

$$\Delta Z = A\Delta H + B\Delta D \tag{4.2}$$

where A and B are constants peculiar to the station (RIKITAKE and YOKOYAMA, 1955).

A and B are in general complex quantities when we apply spectral analysis technique to analysis of magnetograms. They are called the transfer functions. Such a tendency as indicated in Eq. (4.2) is caused by a nonuniform distribution of electric conductivity within the earth. Study of conductivity anomaly (CA) is now flourishing (e.g. HONKURA, 1978b).

In RIKITAKE (1976a) are summarized hitherto-reported changes in A and B values preceding an earthquake. RIKITAKE (1979b) showed that the semi-annual mean values of A exhibited a remarkable change prior to the Sitka, Alaska, earthquake ($M = 7.2$, 1972) as shown in Fig. 4.28. The earthquake occurred at a location only 40 km distant from the magnetic observatory. Much caution was taken in the analysis in order to eliminate the influence of the external magnetic field arising from the auroral electric currents because the geomagnetic latitude of the Sitka Magnetic Observatory is close to that of the auroral zone.

SANO (1978), who extended the work by YANAGIHARA (1972) and YANAGIHARA and NAGANO (1976), showed that the superposition of A and B values for 14 earthquakes in the vicinity of Ibaraki Prefecture having epicentral distances of several tens of kilometers at maximum from the Kakioka Magnetic Observatory indicates clear-cut decreases in their real parts about 20 days prior to the day of earthquake occurrence. The superposition is made in such a way that the day of earthquake occurrence is taken as the time origin. As the amplitude of change in the transfer functions is less than 0.01, however, physical significance of such changes should be carefully examined.

Y. Kato, M. Sato, and T. Hayasaka (personal communication,

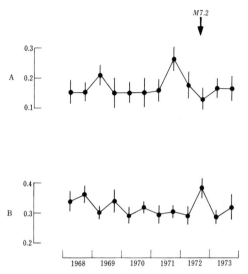

FɪG. 4.28. Secular variation of CA transfer functions A and B at the Sitka, Alaska, Magnetic Observatory.

1979) confirmed that the *A* value at the Onagawa Magnetic Observatory adjacent to the aftershock area of the Miyagi-Ken Oki earthquake (*M* = 7.4, 1978) remarkably increased several months forerunning the quake.

In China, studies on the relation between CA transfer function and earthquake occurrence are going on. Xu *et al.* (1978) showed that they observed a change in *A* value preceding the Songpan-Pingwu earthquakes (*M* = 7.2, 6.7, 7.2, 1976).

4.6.4 Earth potential

Electric potential difference between two electrodes buried in the ground can be easily measured with a rather simple instrument. Earth potential measurement is therefore very popular at amateur stations in China. As a result many reports related to earth potential work at the time of the Haicheng, Liaoning Province, earthquake (*M* = 7.3, 1975) are available.

Figure 4.29 is the change in earth potential observed at the 13th Shenyang Middle School about 145 km distant from the epicenter.

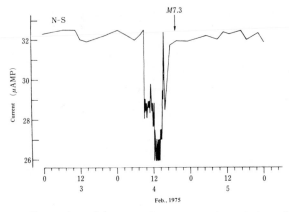

FIG. 4.29.   Observations of the current between a north-south electrode pair (about 50-m separation) at Shenyang Middle School 13 (145 km from the epicenter of the Haicheng earthquake). During the period of the anomaly, readings were made as rapidly as possible, but before 0930 and after 1630 on February 4, only hourly readings are plotted (RALEIGH *et al.*, 1977).

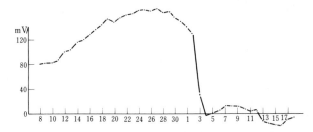

FIG. 4.30.   Daily averages of the self-potential observed over a 70-m (north-south) electrode separation by the 102 Bridgade for Geological Prospecting at a station close to the epicenter of the Haicheng earthquake ($M = 7.3$, Feb. 4, 1975) (ZHU, 1976).

The electrodes are buried in the ground with an interval of about 50 m (RALEIGH *et al.*, 1977).

Figure 4.30 is the change in earth potential as observed by the 102 Brigade for Geological Prospecting at a station very close to the epicenter. The brigade, which consists of 600 members, is good at this kind of observation. When the author called on the brigade, Wang Fu-chang, the chief observer, gave a lecture about the earth potential observation.

The observation was commenced from January 7, 1975 in view of the short-term earthquake prediction. The lead-plate electrodes are buried at a depth of 120 cm. The electrode intervals are 70 and 50 m respectively for the north-south and east-west components. The usual depth of underground water level is 2–3 meters from the ground surface. They observed a large change in the underground water level in January 18.

Observations are made hourly with a transistorized self-compensating potentiometer. It is assumed that the daily value is represented by the average of four readings at 0, 6, 12, and 18 h. Figure 4.30 is the change in daily mean values thus obtained for the north-south component. Changes in earth potential for both the components during a longer period are shown in RALEIGH et al. (1977).

Many pulses began to be observed from around 16 h on February 3, the day before the main shock. It was sometimes difficult therefore, to make an observation for the north-south component. The potential difference between the north and south electrodes exceeded 1 volt around 23–24 h on that day. The earth potential observation seems likely to be disturbed by rain fall. In August of the year anomalous changes of large amplitude were observed as can be seen in the observed results quoted by RALEIGH et al., (1977). There are some more amateur stations where precursory changes in earth potential were observed.

NORITOMI (1978a, b) reviewed many examples of precursory change in earth potential in relation to destructive earthquakes in Yunnan, Sichuan, and Hebei Provinces. Those particularly interested in such examples should refer to NORITOMI (1978a, b).

When the Japanese seismological delegation headed by the author visited Sichuan Province in 1978, we could see some anomalous changes in earth potential that preceded the Songpan-Pingwu earthquakes ($M = 7.2$, 6.7, 7.2, 1976). Figure 4.31 shows the change in earth potential observed at a middle school in Mianzhu about 130 km from the epicenter. Similar changes at Shuijing and Gucheng near the epicenter are shown in Fig. 4.32. The epicentral distances are 20 and 50 km, respectively.

The only U.S. examples of precursory change in earth potential, to the author's knowledge, are reported by CORWIN and MORRISON (1977). A decrease in earth potential amounting to 140 mV/km was

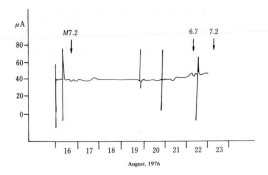

FIG. 4.31.   Earth potential record obtained at a middle school in Mianzhu in relation to the Songpan-Pingwu earthquakes ($M = 7.2$, 6.7, 7.2, 1976).

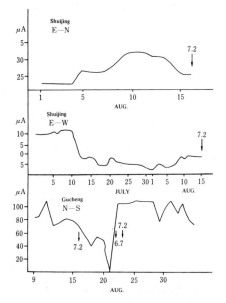

FIG. 4.32.   Earth potential records obtained at Shuijing and Gucheng in relation to the Songpan-Pingwu earthquakes ($M = 7.2$, 6.7, 7.2, 1976).

observed at a station 37 km distant from the epicenter of a magnitude 5.2 earthquake (1974) near Hollister, California 55 days prior to the shock. On the occasion of magnitude 2.4 (1975) earthquake, a 13 mV/km decrease is also observed 4.6 days prior to the shock. The epicentral distance was 2.5 km.

KOYAMA and HONKURA (1978) made an earth potential observation at a station near Sugehiki (see Fig. 4.23). They observed a change that seems likely to precede the Izu-Oshima Kinkai earthquake ($M = 7.0$, 1978) as can be seen in the lower graph of Fig. 4.22. The precursor time and epicentral distance are respectively 64.5 days and 30 km in this case.

4.6.5 Earth resistivity

In RIKITAKE (1976a) are reviewed precursory changes in ground resistivity as observed in the U.S.S.R., the U.S.A., China, and Japan. The aim of this subsection is to add new findings of similar sort to the existing data.

In contrast to the precursory decrease in resistivity in the case of an earthquake ($M = 3.9$) and a swarm activity ($M = 3.5$ in total) on the San Andreas fault in 1973 (MAZZELLA and MORRISON, 1974), no appreciable change in resistivity was observed for the October, 1977 earthquake ($M = 3.8$) and December, 1977 earthquake ($M = 4.0$), the epicenters of these quakes being located not very far from the last shocks that accompanied resistivity changes (MORRISON et al., 1979).

Measurement of earth resistivity as a means of detecting earthquake precursor is very popular in China. ZHAO and QIAN (1978) presented a resistivity change that was brought to light a few years prior to the Tangshan earthquake ($M = 7.8$, 1978) as shown in Fig. 4.33. According to experiments on rock samples from the Tangshan area, the mechanical strain caused by uniaxial compression is amplified 1,000 times if it is measured in terms of resistivity change. The decrease in resistivity as shown in Fig. 4.33 may be accounted for provided a compressional strain of 3×10^{-5} predominates in the focal area.

NORITOMI (1978a, b) summarized many Chinese examples of precursory resistivity change. Among them, the resistivity decrease observed at Changli, about 80 km distant from the epicenter of the Tangshan earthquake, is the most remarkable (see Fig. 8.2). The precursor time amounts to 5 days in this case.

Changes in resistivity is observed at the Chengdu, Sichuan Province, Observatory with a pair of electrodes for electric current having electrode intervals of 736 and 846 m respectively in directions of N57°E and N47°W and a pair of electrodes for picking up potential

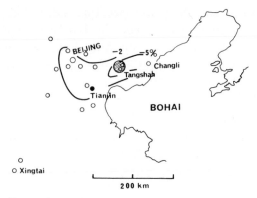

FIG. 4.33. Changes in resistivity as expressed in percentage preceded the Tangshan earthquake ($M = 7.8$, 1976). The shaded and open circles show the epicenter and observation points, respectively (ZHAO and QIAN, 1978).

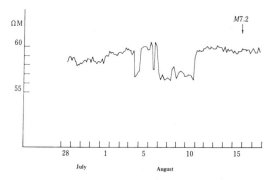

FIG. 4.34. Changes in resistivity at the Chengdu Standard Observatory precursory to the Songpan-Pingwu earthquakes ($M = 7.2$, 6.7, 7.2, 1976).

drop having electrode intervals of 226 and 270 m respectively in the directions cited above. Usually, an electric current of 1 ampere driven by a voltage of 220 V is sent into the ground. Figure 4.34 is the anomalous change in resistivity observed by the observation. The result of a similar observation at Dayi, about 50 km west of Changdu, is shown in Fig. 4.24.

SIDORENKO et al. (1979) reported on changes in earth resistivity preceding earthquake swarms and moderately large earthquakes observed at the Cherkey hydroelectric power plant on Sulak River in Dagestan.

It has often been reported that the highly-sensitive resistivity variometer at Aburatsubo, a near-shore station about 60 km south of Tokyo, registers precursory changes in resistivity having a precursor time of a few hours (YAMAZAKI, 1975; RIKITAKE and YAMAZAKI, 1976, 1977). When a strain of the order of 10^{-9}–10^{-8} is observed in terms of resistivity change, it is amplified 10^4 times or so by a not-entirely-identified mechanism in the formation there.

RIKITAKE and YAMAZAKI (1979) further showed that a precursor having a longer precursor time was observed by the variometer in the case of the Izu-Hanto Oki earthquake ($M = 6.9$, 1974).

In contrast to the outstanding precursory change in resistivity that started 4 hours prior to the Izu-Hanto Oki earthquake, of which the epicentral distance was 100 km, no recognizable precursor was observed for the Izu-Oshima Kinkai earthquake ($M = 7.0$, 1978). The epicentral distance amounted to only 60 km this time. It is not entirely clear why we observe such a difference although it is suspected that the difference in source mechanism has something to do with such contradiction.

4.7 Gravity

TANAKA (1978a) summarized the gravity changes associated with earthquakes in China. Figure 4.35 shows the results of repeated gravity surveys along a route from Beizhen, a city in the middle of Liaoning Province, to Zhuanghe, about 240 km southeast of Beizhen. Askania Gs-11 and Gs-12 type gravimeters are used for the measurement. As can be seen in the figure, in which it is assumed that there was no gravity change at Beizhen, a decrease in gravity amounting to -264 μgals (1 gal = 1 cm/s^2) took place along the southeast half of the route during June to November, 1972. Such a decrease in gravity seems to be extended to the northwest part of the route according to the May, 1973 survey. The largest decrease since the first survey amounts to -352 μgals. According to the survey immediately after the Haicheng earthquake ($M = 7.3$, February 4, 1975), the pre-earthquake decreases in gravity seems likely to have vanished over the whole route.

The gravity then increased as disclosed by a survey in July, 1975, the maximum increase reaching 378 μgals near the southeastern end

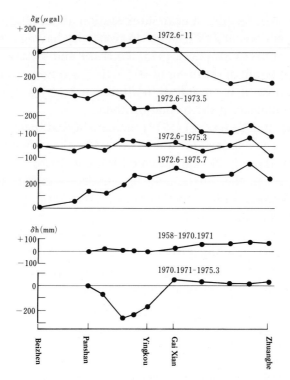

FIG. 4.35. Changes in gravity and height at bench marks along a route in southern Liaoning Province (TANAKA, 1978a).

of the route. The changes in height as revealed by levelling surveys are also shown in Fig. 4.35. The changes in gravity are so large that they cannot be accounted for either by vertical deformations of land or density changes due to compaction, so that it cannot be helped to assume some kind of mass transfer within the earth.

TANAKA (1978a) further showed the gravity changes associated with the Xingtai ($M=6.8$, 7.2, 1966), Heijian ($M=6.3$, 1967) and Tangshan ($M=7.8$, 1976) earthquakes. Comparing the result of gravity measurements in March and April, 1976 to that in June, 1976, the period just prior to the Tangshan earthquake, it appears that an increase in gravity amounting to 150 μgals has taken place in the Tangshan area.

HAGIWARA et al. (1976), HAGIWARA (1977), and GEODETIC SURVEY PARTY (1978) detected a gravity decrease of about 30 and 50

μgals at maximum respectively during December, 1974–March, 1976 and December, 1974–February, 1978 in association with the anomalous uplift as shown in Fig. 4.1. The gravity change thus disclosed seems likely to be accounted for by the decrease of free-air rate due to increase in height. There is no evidence that a mass of high desity has come up close to the earth's surface from below.

According to the experiences in Japan and the U.S.A., no gravity change exceeding scores of microgals has been observed. The reason why they frequently observe extremely large changes in gravity reaching a few hundred microgals in China is not known.

4.8   Underground Water

Very few systematic studies on anomaly of underground water have been conducted in relation to an earthquake in Japan. WAKITA (1978c) presented 113 examples of anomaly of underground water in a summarized form.

SATO (1977) newly pointed out the anomalous change in the level of underground water preceded the Nankai earthquake($M = 8.1, 1946$). These data have been overlooked for a long time. Seventeen examples of lowering of well-water level and/or muddy water were reported from the western Shikoku and the eastern Shikoku-Kii Peninsulas respectively 5–7 and 1–2 days prior to the quake.

In the case of the Izu-Oshima Kinkai earthquake ($M = 7.0$, 1978), YAMAGUCHI and ODAKA (1978) reported a precursor-like change in underground water level at a well about 35 km distant from epicenter as shown in Fig. 4.36.

It turns out that the change in underground water level prior to the Meckering, Australia, earthquake ($M = 6.9$, 1968) (GORDON, 1970), which was quoted in RIKITAKE (1976a), is nothing but an effect caused by rainfall (GREGSON et al., 1976).

WU (1975) reported that the pressure of gas-wells in Taiwan increased 15–16% 10 days preceding an earthquake of magnitude 6.7, the epicentral distance being 20 km. Such a change appeared at 3 wells having a depth of 15 km. No such violent change has ever been observed in the past.

It is quite impossible to deal with all the underground anomalies

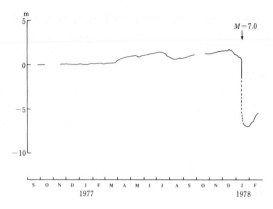

Fig. 4.36.   Changes (corrected for changes in atmospheric pressure) in underground water level precursory to the Izu-Oshima Kinkai earthquake ($M=7.0$, 1978) as observed at a well at Funahara, Izu Peninsula. The well, of which the depth is 600 m, is about 35 km distant from the epicenter (YAMAGUCHI and ODAKA, 1978).

reported in China because they are too numerous. As has already been mentioned in Section 1.4, there are many examples such as the well from which water spouted out before the Haicheng earthquake and the like. They may be classified into the so-called "macroscopic effect" (see Subsection 4.10.1) that can be observed without any instruments.

WAKITA (1978a, b) showed a number of underground water precursors observed in China. Figure 4.37 shows the changes in underground water level precursory to the Tangshan earthquake ($M=7.8$, 1976) observed at the Dahuichang Crustal Movement Observatory, about 180 km west of the epicenter. It is remarkable that the shape of the graph shown in the Fig. 4.37 resembles that of the changes in ground tilting shown in Fig. 4.8.

About 3 months before the Longling earthquakes ($M=7.5$, 7.6, 1976), the water level at a well at Xiaguan, Yunnan Province began to lower by scores of centimeters, the epicentral distance being 180 km. Some 10 days after the water level tended to increase, the earthquake occurred. Some more examples of this kind can be found in WAKITA (1978a, b).

It seems likely that we have to choose suitable wells for detecting precursory changes in underground water. Those wells for irrigation cannot be used for the purpose. When the underground water is

F<small>IG</small>. 4.37. Changes in underground water level precursory to the Tangshan earthquake ($M = 7.8$, 1976) as observed at the Dahuichang Crustal Movement Observatory (W<small>AKITA</small>, 1978a).

running rapidly, it is also difficult to take precursory data. The best thing to do is to dig a deep well only for monitoring precursory changes in underground water.

4.9 Geochemical Precursor

A remarkable changes in radon content was reported in association with the Gasli, the U.S.S.R., earthquake ($M = 7.3$, May 17, 1976) (S<small>ULTANXODJAEV</small> et al., 1976). Figure 4.38 shows the change in radon content observed at Ulegbek near Tashkent. The distance from the epicenter to the observation point amounts to 400 km. The depth of the well from which the sample water was taken was 1,800 m. As can be seen in the figure, the change in radon content that occurred in the evening of May 13 is notable.

On the basis of the above result, a warning of earthquake occurrence was issued on May 15, although the location of epicenter cannot be predicted. This kind of earthquake prediction was issued at least 6 times in the past, while four of them were successful. In order to do this kind of observation, it is required to choose appropriate wells. Only one in four wells is qualified for such an observation.

As has been stated in Subsection 1.2.2(7), a series of measurements of radon content have been conducted along the San Andreas fault. In general, annual change is usually predominating. However, changes that seem likely to have something to do with occurrence of earthquakes of magnitude 4 or so have been noticed.

Measurement of radon content is now getting popular in Japan. Faculty of Science of the University of Tokyo, Geological Survey of

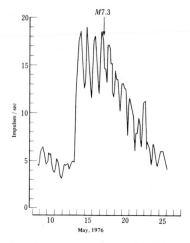

FIG. 4.38. Changes in radon concentration at Ulegbek preceded the Gasli earthquake ($M = 7.3$, 1976) (SULTANXODJAEV *et al.*, 1976).

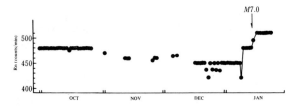

FIG. 4.39. Changes in radon concentration at a well at Nakaizu about 30 km in epicentral distance preceded the Izu-Oshima Kinkai earthquake ($M = 7.0$, 1978) (Faculty of Science, University of Tokyo, unpublished, 1978).

Japan, and Tokyo Metropolitan Office are undertaking such measurements in the Tokai, Sourth Kanto and Tokyo areas. Figure 4.39 is the change in radon content at a well in the Izu Peninsula. It seems likely that the radon content indicated a precursory change preceding the Izu-Oshima Kinkai earthquake ($M = 7.0$, 1978) (unpublished, Faculty of Science, University of Tokyo, 1978).

In China, radon content observation is widely made for the purpose of earthquake prediction. According to WAKITA (1978a, b), the observation is one of the routine work at 17 standard seismological observatories and about 60 seismological brigades. In addition to this there are countless amateur stations guided by experts. Figure 4.40 shows the measuring system of radon content at the Peking

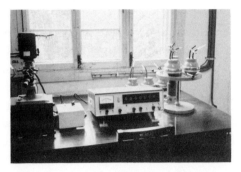

FIG. 4.40. Instruments for measuring radon concentration at the Peking (Beijing) Standard Observatory.

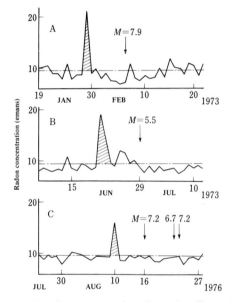

FIG. 4.41. Changes in radon concentration observed at Guzan, Sichuan Province forerunning the Luhuo, Mabian, and Songpan-Pingwu earthquakes (WAKITA, 1978a, b).

(Beijing) Standard Seismological Observatory. The measurement is usually made once a day for the water pumped up from a well having a depth of 158 m.

There are so many reports of precursory changes in radon content in China that it is hard to cover all of them. Those who are particularly interested should refer to WAKITA (1977, 1978a, b). Only the changes observed at Guzan, Sichuan Province, about 210 km southwest of Chengdu are shown here in Fig. 4.41. In this case radon concentration in groundwater issuing from fissures of granite was measured. As can be seen in the figure, spike-like changes were observed prior to the Luhuo ($M = 7.9$, 1973), Mabian ($M = 5.5$, 1973), and Songpan-Pingwu ($M = 7.2$, 6.7, 7.2, 1976) earthquakes. The epicentral distances are 200, 200, and 320 km, respectively. Such a spiky change was also reported at Langfang before the Tangshan earthquake.

According to the Chinese experience, it seems likely that there are two kinds of radon precursor; i.e. long-term precursor and a short-term one having a precursor time of 10 days or so.

In addition to radon measurement, measurement of helium concentration is also conducted in relation to earthquake prediction (CRAIG et al., 1975). SUGISAKI and SHICHI (1978) took gases from a fault in the deepest part of the vault at the Inuyama Crustal Movement Observatory near Nagoya, Japan, and measured N_2/Ar and He/Ar ratios. Measurements were made 1 to 3 times a day. The running averages for 10 days are shown in Fig. 4.42 in which we observe a tendency that He/Ar changes first, then we see a change in N_2/Ar and finally an earthquake occurs. It is speculated that He, which is lighter than N_2, may diffuse faster than N_2 when crustal stress is increased. SUGISAKI (1978) emphasizes the importance of monitoring underground gases and liquids as one of the the means to earthquake prediction.

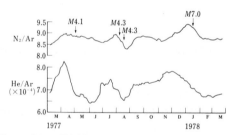

FIG. 4.24.   Fluctuations of N_2/Ar and He/Ar ratios for gases taken from a fault at the Inuyama Crustal Movement Observatory. The epicentral distances of the earthquakes shown in the figure are 100, 15, 45, and 216 km respectively from left to right (SUGISAKI and SHICHI, 1978).

4.10 Macroscopic Anomaly

As has already been mentioned in the preceding chapters, it has often been reported that premonitory anomalous phenomena that can be observed with the eye, ear, and nose sometimes appear forerunning an earthquake. Such anomaly is named by the Chinese the "macroscopic anomaly" in contrast to the "microscopic anomaly" to be observed with sophisticated devices. Although the author feels that the nomenclature in English may not be perfectly correct, the Chinese wording sounds excellent for describing the anomalies.

Data of macroscopic anomaly could be enormous in number in contrast to data taken by geophysical and geochemical instruments which are necessarily limited in number. In this context, macroscopic data can sometimes play a key role in earthquake prediction as has often been reported from China even if the quality of individual data is not high.

Judging from the author's experience during his visit as the chief delegate of the 1978 Japanese seismological mission to China, observation of macroscopic anomalies could be one of the means of earthquake prediction in a country like China where man-made noise is extremely low. There is a Chinese text entitled "Macroscopic Anomaly and Earthquake" (SEISMOLOGICAL BUREAU OF ANHUI PROVINCE, 1978).

It is doubtful, however, that macroscopic anomaly can be observed in a highly industrialized area such as Japan as distinctly as in rural districts in China because of mechanical and electric noise from railways, factories and other sources.

4.10.1 Macroscopic anomaly of underground water

The anomaly of underground water preceded the Haicheng earthquake ($M = 7.3$, 1975) has been mentioned in Section 1.4. In relation to the earthquake, it is reported that water spouted out from two wells in the Panjin area a few days prior to the quake and that water mixed with gases erupted out in a pond at Xinyan breaking the surface ice.

According to WAKITA (1978a, b), the following anomalies of underground water were observed before the Tangshan earthquake

($M = 7.8$, July 28, 1976). A well at a commune in Fengnan, about 5 km distant from the epicenter, spouted up water with an astonishingly loud sound in the early morning on July 28. When they hurriedly woke up all the commune people, the great earthquake took place. The well is an ordinary one having a 100 m depth.

A well of 50 m in depth at a commune in Fengrun, about 18 km distant from the epicenter, started to self-spout about half a month before the earthquake. On July 25 and 26, the volume of spouting water increased along with occasional gas eruptions accompanied by ejection of sand and gravel. The sound associated with the eruption could be heard at a distance of 20 m or more.

Petroleum of about 1 ton spouted up to a height of 2 m from a disused well in the oil field in Maozhou, Hebei Province, about 200 km distant from the epicenter and 140 km south of Peking (Beijing), in the afternoon of July 17.

Figure 4.43 indicates histograms of precursor time for macroscopic anomaly of underground water and of animal (rat in the present case) behavior in regard to the Haicheng and Tangshan earthquakes, respectively. It should be noticed that the peak of precursor time is shifted to a value for the Tangshan earthquake smaller than that for the Haicheng earthquake.

Much of macroscopic anomalies precursory to the Songpan-Pingwu earthquakes ($M = 7.2$, 6.7, 7.2, August 16, 22, 23, 1976) came to author's knowledge when the author and his mission visited Sichuan Province in 1978. Zhu Jie-zuo of the Sichuan Seismological Bureau gave a lecture on the macroscopic anomalies at a meeting in a hotel in Chengdu (see Fig. 4.44).

According to Zhu, macroscopic anomalies started to be observed in an area of the Longmen Shan fault system including Dayi to the southwest of Chengdu in June, 1976. They are mainly anomalies of underground water and small animals. The area is designated as area 1 in Fig. 4.45. In the meantime macroscopic anomalies tended to appear in area 2. Towards the end of July and the beginning of August, the reports on macroscopic anomaly increased enormously in area 3 which is close to the epicenter.

Figure 4.46 is the picture of a fountain in Jiangyou of which the epicentral distance was about 100 km. The water level of the fountain

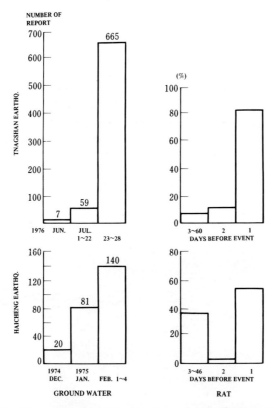

FIG. 4.43. Time sequence of the occurrence of macroscopic precursor such as changes in underground water and anomalous behavior of rats. Cases for the Tangshan and the Haicheng earthquakes are compared (WAKITA, 1978a).

FIG. 4.44. A lecture on macroscopic anomalies precursory to the Songpan-Pingwu earthquakes by Zhu Jie-zuo of the Sichuan Seismological Bureau.

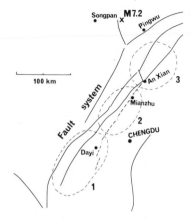

FIG. 4.45. Three areas over which macroscopic anomalies are successively observed prior to the Songpan-Pingwu earthquakes.

FIG. 4.46. Premonitory lowering of water level at a fountain in Jiangyou before the Songpan-Pingwu earthquakes (After the Sichuan Seismological Bureau).

FIG. 4.47. A giant panda died of hunger because of the death of bamboo due to lowering of underground water-head (After the Sichuan Seismological Bureau).

lowered towards the beginning of August. Such a lowering of the head of underground water gave rise to the death of bamboo, so that giant pandas died of hunger as can be seen in Fig. 4.47. They eat only leaves of bamboo of a particular kind.

4.10.2   Earthquake light and fire ball

According to Sun Ya-jie, the vice-leader of the Dingjiagou Production Brigade, who experienced the Haicheng earthquake ($M = 7.3$, 1975) right in the focal area, a light flashed on the ground about 30 m distant from her 0.5 second before the shock. She later found a crack there. It is also said that a line of light appeared in a direction of the nearest mountain where an earthquake fault was found later.

The drivers of a locomotive engine on the night train from Peking (Beijing) to Dalian saw three lightenings in front of them when they were approaching Tangshan. They operated the brake immediately. When the train stopped, the great Tangshan earthquake ($M = 7.8$, 1976) took place.

These reports suggest that earthquake light is observed almost at the time of earthquake occurrence. However, we were told by the member of the Sichuan Seismological Bureau that a reddish rainbow-like light with purple color appeared very high in the sky at Kangding at 21 h 05 m on August 15, 1976, one day before the Songpan-Pingwu earthquakes. The epicentral distance was 350 km.

While those reports on earthquake light do not clearly describe that the anomalous light is a kind of earthquake precursor, Zhu reported on fire balls which are certainly a kind of precursory effect. He himself could see one of them before the Songpan-Pingwu earthquakes ($M = 7.2$, 6.7, 7.2, 1976). As many reports on fire balls that came out from the ground reached the Sichuan Seismological Bureau some time in July, 1976, about one month preceding the main shocks, Zhu went to the Mianzhu area where he could see a fire ball in the evening on July 19. According to him a fire ball as large as a washbowl appeared from the ground. The fire went up to the sky to about 10–20 m in height. When it vanished, the size of the fire ball was something like a ping-pong ball. The color was a mixture of red, blue, and white. It is said that leaves were burnt when a fire ball touched them. The fire balls were associated with odours similar to garlic or sulphor, so that there were people who felt ill and nauseate from such odours.

Most fire balls were observed at an intersection area of the Longmen Shan fault system and rivers, the diameter of the area being 5–6 km. Towards the middle and end of July, there were about 150 reports on fire balls. It is said that an explosive sound is associated with fire ball appearance.

As it is understood that they have plenty of natural gases in the Sichuan basin, fire balls can possibly be accounted for by burning of natural gases emitting from the ground because of increasing crustal stress.

4.10.3 Gas spouting

Probably, it is better to treat the fire-ball phenomenon in this subsection. There are many instances for which gas spouting is observed before an earthquake.

It was reported that the water of a well in the focal area of the Haicheng earthquake bubbled and became muddy about one month prior to the quake.

As can be seen in Fig. 4.48, the water in a pond in Jiangyou bubbled about 20 days preceding the Songpan-Pingwu earthquakes. According to the result of chemical analysis, the composition of the spouting gas was as follows; N_2 94%, CO_2 0.06%, and CH_4 0.95%.

FIG. 4.48. Bubbling in a pond in Jiangyou precursory to the Songpan-Pingwu earthquakes (After the Sichuan Seismological Bureau).

FIG. 4.49. Small upheavals in paddy fields prior to the Songpan-Pingwu earthquakes (After the Sichuan Seismological Bureau).

About the same time, people found small upheavals in paddy fields in Mianzhu as shown in Fig. 4.49. There was a hole on top of the upheavals. It is supposed that natural gases came out from the hole.

The vice-chairman of the revolutionary committee at Pingwu town confirmed the bubbling at a fish-farming pond one day before the earthquake.

4.10.4   Anomalous behavior of animals

As was reviewed by RIKITAKE (1976a), there have been many reports on anomalous behavior of animals before an earthquake occurrence. It is true, however, that those reports are in many cases ambiguous and fragmental, so that most seismologists did not treat them as scientific data until recently.

In recent years, however, many convincing reports on precursory animal behavior circulated, following the Haicheng, Tangshan, and Songpan-Pingwu, China, earthquakes, Friuli, North Italy, earthquake ($M = 6.7$, 1976), and Vrancea, Romania, earthquake ($M = 7.2$, 1977). Under the circumstances, anomalous animal behavior became one of the disciplines of earthquake prediction study, and so budgetary funds have been offered to researches dealing with animal behavior in Japan as well as in the U.S.A.

USGS convened a symposium on anomalous behavior of animals aiming at summarizing the existing knowledge (EVERNDEN, 1976). RIKITAKE (1978d) published a popular book entitled "Can

Animals Predict an Earthquake?" in which many existing data including those found in historical documents in Japan and other countries are summarized. SEISMOLOGICAL PARTY, BIOPHYSICAL INSTITUTE OF THE ACADEMIA SINICA (1977) also published a book entitled "Animal and Earthquake".

No individual example of animal behavior precursory to an earthquake will be given here. Those who are particularly interested in animal behavior related to an earthquake should refer to the above and following literature. RIKITAKE (1978b), who applied a Weibull distribution analysis to the existing 157 data, i.e. 29, 27, 90, and 11 for mammals, birds, fish and snakes, frogs, insects and earth worms, respectively, analysed the precursor time of animal behavior. He found two peaks of precursor time distribution, one is around 2–3 hours and another around half a day. The latter peak covers the time range for which geophysical and geochemical precursors are seldom observed, so that, should it be a precursor at all, animal precursor is complementary to geophysical and geochemical precursors.

RIKITAKE (1979a, 1981b) and RIKITAKE and SUZUKI (1979) conducted a study of animal anomaly respectively in relation to the Izu-Oshima Kinkai ($M = 7.0$, 1978) and the Miyagi-Ken Oki ($M = 7.4$, 1978) earthquakes by sending enquiry cards to schools, town and city offices, zoos and the like in the earthquake-shaken areas. Figure 4.50 shows the locations where animal anomaly is observed for the former quake. It is clear that the smaller the epicentral distance is or the larger the earthquake damage is, the larger is the number of reports on animal anomaly. On the basis of 129 data thus collected, the histogram of precursor times of animal anomaly is shown in Fig. 4.51 in which we can see two peaks around scores of minutes and 0.6 day. No remarkable difference in precursor time between animals of different species is found. The distribution as shown in Fig. 4.51 is not much different from that for the data set of animal anomaly reported in the past. The analysis of the data for the Miyagi-Ken Oki earthquake leads us to a more or less similar conclusion although the second peak occurs around 0.3 day in this case. In any case these studies did not reach a conclusion that animal anomaly has nothing to do with earthquake occurrence.

The author drew a histogram of precursor time of animal

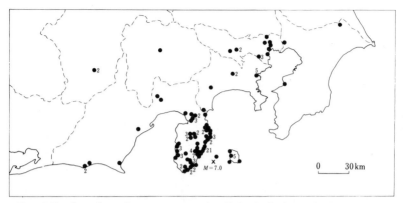

FIG. 4.50. Locations where precursory animal behavior is observed prior to the Izu-Oshima Kinkai earthquake ($M = 7.0$, 1978), the epicenter of which is marked by a cross. The numerals indicate the number of reports. When no numeral is attached, only one report is available from that location.

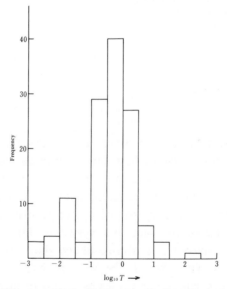

FIG. 4.51. The histogram of logarithmic precursor time T (unit: days) for the whole data of animal behavior precursory to the Izu-Oshima Kinkai earthquake.

anomaly by making use of all the data available to him at present. As can be seen in Fig. 4.52, the peak around 0.5 day is conspicuous. Relatively few data from China are involved in the analysis.

When the author and his seismological mission visited Sichuan

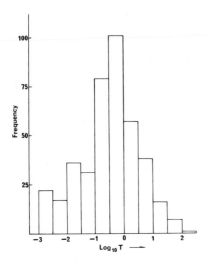

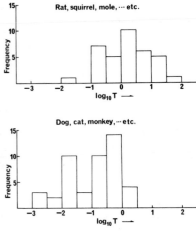

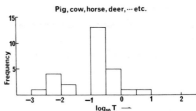

FIG. 4.52. The histogram of logarithmic precursor time T (unit: days) for all the available data of animal behavior.

FIG. 4.53. The histogram of logarithmic precursor time T (unit: days) for mammals classified according to the size of body.

Province, they were told by Chinese colleagues that small animals become tumultuous earlier than large animals. Whether such a tendency is true or not is checked by the data set used for the analysis of Fig. 4.52. Only data for mammals can be used for the analysis because it is hard to judge the size of birds and fish only by the name of species. In spite of the fact that very few data from China are involved in the analysis, it seems likely that what our Chinese colleagues said is true. As can be seen in Fig. 4.53, the shift of precursor time towards larger values as we go to large animals from small ones is evidently observed in the figure in which histograms of precursor time for three classes of animals: small—rat, squirrel, mole etc.; medium—dog, cat, monkey etc.; large—pig, cow, horse, deer etc., are shown, respectively.

It is not entirely clear at the moment what kind of premonitory effect stimulates animals. Should the physical causes that excite animals become clear, we geophysicists would monitor direct that stimulus. It is urgently hoped, therefore, that the physical causes that excite animals are identified by biologists.

One of the serious defects of the study of animal behavior is the point that most data are not quantified. Even if one says that an animal of a particular species behaves anomalously, it is totally subjective, so that no quantitative comparison between data taken at different places and times can be made. Attempts for such quantification are most advanced in the U.S.A. as reported in EVERNDEN (1976). For instance, the frequency for a chimpanzee passing through a door, that connects its usual living place to the next room, is counted. Such a frequency seems likely to represent chimpanzee's restlessness. It is reported by Stanford University that an earthquake swarm occurred one day after observing unusually high restlessness. Similar effort towards quantification has further been made for other species in the U.S.A. and China in recent years although no more detail is given here.

4.10.5   Other macroscopic anomalies

It is sometimes said, especially in China, that even a plant exhibits anomaly preceding an earthquake. For example, the flower of a certain tree comes into bloom while the tree bears fruit before an earthquake. It is also said that it turns sultry immediately before an earthquake. Contradictory to this, there is a saying that an earthquake will occur when it turns cold in summer.

It is hard for the author to believe those precursory anomalies of plant and weather because fluctuations of these phenomena are large. Generally speaking, we cannot think of any physical reasons that connect these anomalies to earthquake occurrence.

In China, earth sounds without ground shaking and noise of radio waves are regarded as a kind of premonitory effect. Although the possibility of these precursory phenomena cannot be ruled out, it is hard to say that they are established on the basis of ample data. Much of precursory phenomena of this kind is described in SEISMOLOGICAL BUREAU OF ANHUI PROVINCE (1978).

4.11 Classification and Characteristics of Earthquake Precursor

RIKITAKE (1975c) examined the relationship between precursor time T and earthquake magnitude M in Richter scale of main shock for then-available data amounting to 282 in number. The study was later extended to a data set for 391 geophysical and geochemical precursors (RIKITAKE, 1979d). Those precursors are listed in Table 4.1.

TABLE 4.1. Number of precursors.

Discipline	Abbreviation	Number of data
Land deformation	l	30
Tilt and strain	t	89
Tidal strain	t'	2
Foreshock	f	83
Anomalous seismicity	f'	14
b-value	b	12
Microseismicity	m	4
Source mechanism	s	7
Hypocentral migration	h	4
Fault creep anomaly	c	2
V_P/V_S	v	50
V_P and V_S	w	19
Geomagnetism	g	6
Earth currents	c	17
Resistivity	r	32
Conductivity anomaly	c'	3
Radon	i	12
Underground water	u	2
Oil flow	o	3
	Total	391

As it appears that there is no relation between T and M for foreshock, ground tilting observed with a horizontal pendulum tiltmeter, earth-currents and the like, a $\log_{10} T$ versus M diagram for the precursor data, from which the data of the said disciplines are excluded, is drawn as shown in Fig. 4.54. T is measured in units of days. It is clearly seen in the figure that there are two groups of precursor; the one for which $\log_{10} T$ is proportional to M and the other scattering around $T = 0.1$ day irrespective to M. Let us call them the precursor of the first kind and of the second kind, respectively.

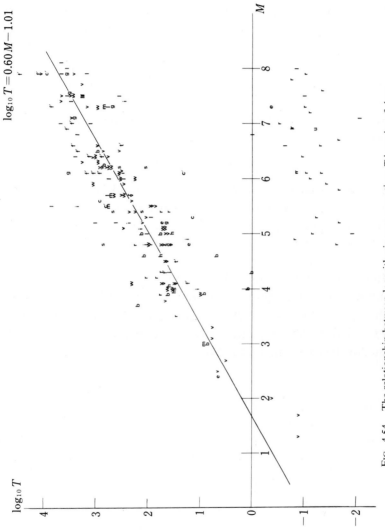

FIG. 4.54.   The relationship between logarithmic precursor time T in units of days and earthquake magnitude M of main shock. The alphabets correspond to those in Table 4.1 and show the disciplines of geophysical and geochemical precursor.

On making use of the least squares method, the best fitting straight line for precursors of the first kind is obtained as

$$\log_{10} T = 0.60M - 1.01 \qquad (4.3)$$

and the straight line expressed by the relation is shown in Fig. 4.54.

It is interesting to note that the $\log_{10} T - M$ relation as obtained here roughly agrees with those obtained by TSUBOKAWA (1973) and SCHOLZ et al. (1973) who analysed data of which the sample number is smaller by one order of magnitude than that for the present analysis.

According to Ding Guo-yu (personal communication, 1978), an analysis of the same kind for Chinese data leads to

$$\log_{10} T = 0.38M - 0.34 \qquad (4.4)$$

It may be that the $\log_{10} T - M$ relation could be different from region to region according to seismotectonic conditions.

The mean precursor time of the precursors of the second kind, which are scattering around $\log_{10} T = -1$, is estimated as 2.4 hours with a standard deviation of 2.3 hours.

Figure 4.55 represents a histogram of precursor times of all disciplines, from which those belonging to the first kind are omitted though foreshocks, tilts, earth-currents and the like, which have been excluded in the foregoing analysis, are included this time. Only earthquakes having a magnitude of 6 or over are dealt with. We see

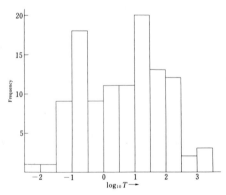

FIG. 4.55.   The histogram of logarithmic precursor time in units of days of the third-kind precursors.

two peaks in the figure around $\log_{10} T = 1$ and -1, respectively. The latter peak coincides with the precursors of the second kind.

A Weibull distribution analysis indicates that the mean logarithmic precursor time and its standard deviation for the former group amount to 0.65 and 1.1, respectively. This corresponds to a mean precursor time of 4.5 days although the distribution spectrum is fairly broad. We may call this class of precursors the precursors of the third kind.

4.12 Physical Cause of Earthquake Precursor

No definite reason why we sometimes observe a precursor before an earthquake has been worked out. It has been proposed, however, that a dilatancy model (e.g. SCHOLZ et al., 1973) could account for most effects precursory to an earthquake. According to the model, the process in the earth crust that leads to occurrence of an earthquake is as follows.

When the crustal stress increases to some extent, many microcracks are to be produced in the rocks composing the crust. This state is called the dilatancy which necessarily leads to an increase in volume. In that case, it is quite natural to expect an upheaval at the ground surface. Anomalous uplift often preceding an earthquake could therefore be accounted for as a result of dilatancy generation. It is known from laboratory experiments that the velocity of P wave being propagated through such a dilatant region becomes small although the velocity of S wave is affected very little. It is thus possible to understand the lowering of V_P and V_P/V_S ratio as mentioned in Subsection 4.4.8 on the basis of a dilatancy model.

When underground water tends to fill the newly-opened cracks, the strength of the dilatant region recovers again. This state is called the dilatancy-hardening. In that case, quieting of seismic activity and lowering of b value as discussed in Section 4.4 could be logically explained. It is also expected that electric resistivity in a dilatant region lowers because of water being diffused from the surrounding. Such movements of water through fresh cracks may accompany an increase in radon concentration in underground water. The above series of events may well be expected to take place before the earthquake occurrence which is the result of a fault slip in the dilatant region. It may be said, therefore, that many earthquake precursors can possibly be accounted for, at least qualitatively, provided a dilatancy process is assumed.

RIKITAKE (1975d) demonstrated that the Dambara formula on the relation between earthquake magnitude and area of crustal deformation associated with the earthquake as will be indicated in the following chapter (DAMBARA, 1966, 1979) may be approximately accounted for by introducing a dilatancy model. This is also the case for the Utsu-Seki formula (UTSU and SEKI, 1955) on the similar relation concerning the extent of aftershock area.

MJACHKIN et al. (1975), MOGI (1974), STUART (1974), and BRADY (1974) proposed a dry model of dilatancy which is not necessarily related to water diffusion. In this case the appearance mode of precursors would be somewhat different from that for the wet model as described in the above.

There is no guarantee, however, that earthquakes are always associated with dilatancy generation. As stated in Subsection 4.4.8, it has now become clear that there are many earthquakes that are not preceded by a change in V_P or V_P/V_S ratio. The idea that the dilatancy model could be applied to almost all earthquakes should be put aside.

It has been attempted to explain an anomalous uplift by assuming an aseismic pre-slip, especially in relation to subduction zone earthquakes. But it is not entirely known whether or not such a hypothesis is supported by the data of actual observation. ·

It should be emphasized, however, that the empirical relation for precursors of the first kind such as expressed by Eqs. (4.3) and (4.4) can be approximately accounted for by taking the dilatancy model into account. On the contrary, nothing certain is known about the mechanism of precursors of the second and third kinds. It is only understood that some preliminary rupture begins to take place when precursors of the second kind break out. Precursors of the third kind may reflect the highly strained state of a focal region.

CHAPTER 5

STRATEGY OF EARTHQUAKE PREDICTION

As readers will see in Chapter 13, the "Large-scale Earthquake Countermeasures Act" had been enacted in Japan in December, 1978. Designation to the "Area under Intensified Measures against Earthquake Disaster" was actually made for the 170 cities, towns, and villages in the Tokai area based on the law in August, 1979. Should an imminent prediction by the "Prediction Council" reach the Prime Minister via Director General of JMA, he will issue an "Earthquake Warnings Statement" and a "National Headquarters for Earthquake Disaster Prevention" will immediately be set up.

It is therefore clear that some sort of nation-wide consensus, that earthquake prediction of long- and short-term is possible to some extent, has been established. In spite of this, the scientists including the author who work on earthquake prediction are not quite confident of successful prediction for all destructive earthquakes. In other words, strategy of earthquake prediction has not been completed yet. Such a strategy could be different from country to county.

It is the aim of this chapter to review the possible strategy of earthquake prediction adopted by CCEP although the author's personal view will appear from place to place. A probabilistic approach to earthquake prediction that is not biased by subjective judgment will also be mentioned.

Strategies in other countries will also be reviewed, though briefly.

5.1 Areas for Special, Intensified and Concentrated Observation

On February 20, 1970 CCEP designated the South Kanto area an "area of intensified observation" and 8 other areas of Japan to "areas of special observation" as already mentioned in Table 1.2. The reasons of designation to an area of special observation vary from area to area. They are (1) area that experienced a large earthquake in the past, (2) area in which an active fault is found, (3) area in which earthquakes occur frequently, and (4) area which is very important socio-economically like Tokyo, Osaka, and so on. As an anomalous ground uplift was observed on the Boso Peninsula, the South Kanto was designated an area of intensified observation. The Tokai area was subsequently designated an area of intensified observation in 1974 because it was confirmed that an accumulation of crustal strain is enormous there.

The strategy proposed by CCEP is as follows. First of all, geodetic surveys of nation-wide scale should be intensified along with more intensified surveys and construction of observatories in areas of special observation. When anomalies are found by the surveys and observations, the area is to be designated to an area of intensified

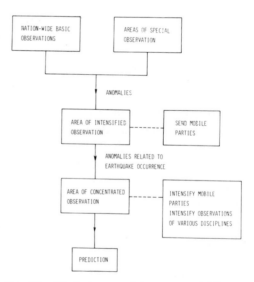

Fig. 5.1. Block-diagram of the strategy of CCEP.

observation. Mobile parties will be sent to the area in order to promote geophysical and geochemical observations. Should it be concluded that the anomalies are likely to be connected to occurrence of an earthquake, the area would be nominated to an "area of concentrated observation". Observations of all sorts will then be concentrated on the area until earthquake prediction is finally made. The strategy stated here is illustrated in a diagram in Fig. 5.1.

The strategy in the last paragraph and in Fig. 5.1 sounds just fine so far as the document is concerned. It should be borne in mind, however, that the implementation of such a strategy involves various difficult problems. It is first of all hard to designate a particular area to one of the areas mentioned in the above on the basis of utterly objective judgment. Scientifically speaking, a designation should be made when the probability of an earthquake occurring in a particular area exceeds a previously-prescribed value. It is at the moment difficult to rely on such an objective analysis. It is highly likely that evaluation of probability of an earthquake occurrence is different from member to member of CCEP. The final judgment must, in many cases, rely on how the chairman of the committee evaluates the warning information.

In the earlier period of CCEP, the committee could discuss earthquake risk fairly frankly because the public did not pay much attention to the committee. The public having now recognized the importance and real power of CCEP, the committee is forced to behave with circumspection because it is quite possible that what CCEP says may give rise to social and economic unrest in a certain locale.

In spite of the above-mentioned defects, no realistic strategy of earthquake prediction other than the CCEP's approach can be found because no established way of evaluating earthquake risk is available. It is not practicable to watch the whole territory of Japan with an equal diligence since financial resources and man power are limited. CCEP revised the designation in 1978 basing on the results of observation in recent years. The revised areas of special observation and the two areas of intensified observation are shown in Fig. 5.2. The reason why an area is designated to the area of special observation varies from area to area as summarized in the following.

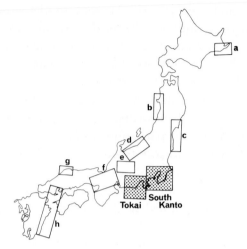

Fɪɢ. 5.2.   Eight areas of special observation and two areas of intensified observation, which are shown by shaded rectangles, are revised by CCEP in 1978.   The reasons for designation of each area are given in the text.

a)   Eastern Hokkaido. The Nemuro-Hanto Oki earthquake ($M = 7.4$) occurred in a seismic gap off East Hokkaido in 1973. However, the crustal deformation, that was prevailing on land, did not vanish and is still there.

b)   Western Akita and north-western Yamagata Prefectures. A number of large earthquakes of magnitude 7 or so occurred in the area in historical time. It appears that seismic activity has tended to increase in the area in recent years. A tilting motion with a southeast-ward dipping has been observed in the Oga Peninsula.

c)   Eastern Miyagi and eastern Fukushima Prefectures. Great earthquakes occur along the Japan trench off the pacific coast of the Sanriku district (Aomori, Iwate, and Miyagi Prefectures). Off the Pacific coast of Miyagi and Fukushima Prefectures, many earth-quakes having a magnitude of 7 or thereabout often occur. A tendency that the seismic activity migrates to east or south directions has been noticed there. A seismic gap is observed in the area, too.

d)   Southwestern Niigata and Northern Nagano Prefectures. Historical evidence has shown that a number of magnitude 7 earthquakes occurred in this area. There are many active faults and foldings along River Shinano. The Niigata earthquake ($M = 7.5$) occurred in a neighboring area in 1967.

e) Western Nagano and Eastern Gifu Prefectures. Many active faults are distributed highly densely in this area. We have had the Fukui ($M = 7.3$, 1948), North Mino ($M = 7.0$, 1961), and Gifu Chubu ($M = 6.6$, 1969) earthquakes in the vicinity of this area in recent years. It appears that seismic activity has been increasing in the area.

f) Nagoya-Kyoto-Osaka-Kobe area. Many earthquakes having a magnitude of 7 or so occurred in the area according to historical evidence. Active faults are concentrated with a high density. A fairly large strain of the earth's crust has been accumulated along the Yoro fault. An anomalous ground tilting with a southward dipping has been observed near the western coast of Lake Biwa. It is obvious that the area is important in socioeconomic terms.

g) Eastern Shimane Prefecture. In the vicinity of this area, we had Hamada ($M = 7.1$, 1872), Tottori ($M = 7.4$, 1943), and Tango ($M = 7.5$, 1927) earthquakes. A large earthquake occurred in the area in documented historical time. To the east of Sanbesan Mountain, a secular uplift has been taking place since the end of the last century. Recently, seismicity becomes high around the mountain.

h) Iyonada Sea and Hyuganada Sea. Seismic activity of magnitude 7 is conspicuously high in this area. There is a tendency that high activity recurs in the area with a return period of 30–40 years. A northward dipping of the ground has been observed on the eastern coast of Kyushu.

The CCEP's designation seems to be acceptable for most scientists working on earthquake prediction. It would be more convincing, however, if the designation is made on the basis of more quantitative specification. As mentioned in Chapters 2 and 3, the probabilities of a great earthquake recurring in areas such as South Kanto, Tokai, and off East Hokkaido can be tentatively estimated. Such probabilities play some role on the designation of areas of special observation. As for the South Kanto area, where the probability of a great earthquake to recur is not high at the moment, the probability of one having a magnitude 6 to 7 occurring right underneath the area is suspected to be fairly high.

It is in most cases impossible to estimate probability of occurrence of an inland earthquake having a magnitude 7 or smaller because the area over which precursory effects can be observed is not wide enough to be monitored by the existing geodetic network. Until the precise geodetic networks as proposed in the 4th 5-year program on earthquake prediction (see Subsection 1.1.2 (1)) are completed, no reliable estimate of occurrence probability of an inland earthquake can be effected except for a few areas where preliminary networks have been finished.

In order to see the risk of inland earthquakes in Japan, the cumulative seismic energy which is released from each prefecture is calculated as shown in Fig. 5.3. As the area is different from prefecture to prefecture, the cumulative energy is divided by the area of prefecture, so that the ordinate of the graph in Fig. 5.3 shows the density of released energy. From the graph one can see both the total energy density since the beginning of Japan's history and that after the year 1600. The latter graph must be more accurate than the former because of preciseness of data. It is assumed that the energy of an earthquake, of which the epicenter is located close to the border of

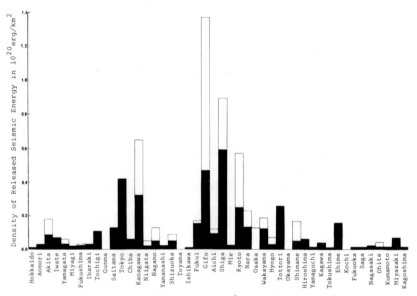

FIG. 5.3.   Cumulative seismic energies per km² released from respective prefectures in Japan. The energies during the whole history are shown with blank columns, while the black portions of the columns indicate the energies released since 1600.

neighboring prefecture, came out from the prefecture that includes the epicenter. Consequently, the graph such as shown in Fig. 5.3 would become a little more accurate if the graph is drawn for a group of neighboring prefectures.

It is observed in Fig. 5.3 that the density of released energy is fairly high for Gifu, Shiga, Tokyo, Kanagawa, Kyoto, Akita Prefectures, and so on. Such a fact is in general harmony with the designation of the areas of special observation by CCEP. As no earthquake energy, which is sent out from the sea area of Iyonada, Hyuganada, and off Miyagi, and Fukushima Prefectures, is involved in Fig. 5.3, however, the energy density thus estimated represents only that for inland earthquakes.

5.2   Synthesized Probability of Earthquake Occurrence

RIKITAKE (1969) tried to formulate the strategy of earthquake prediction as proposed by CCEP in a quantitative form. The main idea is that the estimate of probability of an earthquake of certain magnitude occurring within a certain area during a specified period would become accurate as the number of observed precursors increases. The outline of Rikitake's approach has been quoted in RIKITAKE (1976a), so that no detailed account will be repeated here.

Starting from a preliminary probability, which may be calculated on the basis of routine seismic observation, of an earthquake to take a magnitude value in a certain range, the probability of an earthquake falling in a specified magnitude range can be re-estimated when a precursory land deformation should be observed.

DAMBARA (1966, 1979, 1981), who examined the relation between the extent of crustal deformation associated with an earthquake and the magnitude of main shock, obtained a graph as shown in Fig. 5.4. The best fitting straight line as determined by means of the least-squares method is expressed by

$$\log_{10} r = 0.51 M - 2.26 \qquad (5.1)$$

or

$$M = 1.96 \log_{10} r + 4.43 \qquad (5.2)$$

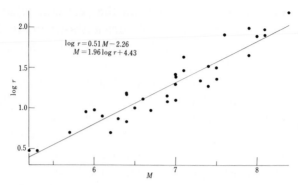

FIG. 5.4. The relation between mean radius of anomalous crustal deformation r (in units of kilometers) associated with an earthquake and magnitude of the main shock M (DAMBARA, 1979, 1981).

where r is the averaged radius in units of kilometers of the anomalous area and M the magnitude of main shock.

As Eqs. (5.1) or (5.2) are only empirical formulas, the magnitude of a coming earthquake can only be estimated probabilistically. It is possible to estimate the probability of the magnitude falling in a certain range by the theory of error based on the standard deviation due to the scatterings as can be seen in Fig. 5.4. The detail of the theory may be found in RIKITAKE (1969, 1976a).

Dividing the whole magnitude range into n intervals, let us denote the probabilities for an earthquake to fall in respective magnitude ranges by $W_1(1)$, $W_1(2), \cdots, W_1(n)$. If the preliminary probabilities for the same magnitude ranges have somehow been obtained as $W_0(1)$, $W_0(2)$, $\cdots$, $W_0(n)$, the synthesized probabilities are given as

$$W(s) = \frac{W_0(s)W_1(s)}{W_0(1)W_1(1) + W_0(2)W_1(2) + \cdots + W_0(n)W_1(n)} \quad (1 \leqq s \leqq n). \quad (5.3)$$

If we can further have precursors of different discipline from which probabilities W_2, $W_3, \cdots, W_k$ for respective intervals can also be estimated, Eq. (5.3) can be generalized as

$$W(s) = \frac{W_0(s)W_1(s)W_2(s) \cdots W_k(s)}{\sum_{m=1}^{n} W_0(m)W_1(m)W_2(m) \cdots W_k(m)}. \quad (5.4)$$

It is therefore seen that the probabilities of an earthquake falling in a specified magnitude range can be estimated on the basis of data of a number of disciplines provided the available data are good enough. It is possible to have a fairly high value of synthesized probability even if the probability for each discipline is not very high. In an actual application, however, only geodetic data are utilizable in almost all cases.

Probability of occurrence time can also be estimated on the basis of an empirical equation such as Eq. (4.3) and preliminary probability which is estimated from a statistical consideration. The logic is more or less the same as that for estimating the probability of magnitude.

When the distributions of magnitude and time of occurrence are assumed to be independent from one another, the probability of having at least one earthquake, of which the magnitude falls in a range from M to $M+dM$, within the period from t to $t+dt$ is given by

$$P(t, t+dt; M, M+dM) = P(t, t+dt)P(M, M+dM) \quad (5.5)$$

in which notations are self-explanatory.

It is thus possible to draw equal probability curves on the magnitude-time plane. Such curves may provide some clue for quantitative designation of areas of special and intensified observations on the condition that estimate of probability could be performed on the basis of a well-organized set of data of high quality. It is regrettable that observations of various disciplines have not been developed to a level that they are always able to be applied to such an estimate of probability, so that the designation of areas of special and intensified observation has been made rather intuitively with the final decision of CCEP's chairman.

5.3 Utsu's Theory on Probabilities of Prediction and Success

It is assumed in the probability theory developed in the last section that an anomalous phenomenon is always connected to an earthquake occurrence. As this is not always the case in nature, UTSU (1977, 1979) tried to introduce two quantities, i.e. probability of prediction and probability of success. Utsu's approach can be applied to estimating the probability of success when a prediction is actually issued and the effect of predition. In Utsu's theory, nothing about earthquake magnitude is included, so that estimate of magnitude must be done on other basis.

Utsu defined the following two probabilities;

p_1 (probability of success): Probability of an anomaly representing an earthquake precursor when it is observed or, in other words, probability of the prediction based on an anomaly being successful.

p_2 (probability of prediction): Probability of an earthquake being preceded by an anomaly or, in other words, probability of an earthquake being predicted.

In this case, we have to define what an anomaly and an earthquake are. They are to be defined by prescribed criteria.

5.3.1   Effect of earthquake prediction

The total amount of earthquake damage when a prediction is issued is denoted by l, while that when no prediction is made by L. We shall further define f and C; f is the expense for disaster prevention that has to be raised as a result of prediction and C the expense spent for observations for earthquake prediction over a certain period of time.

In order to see the effect of earthquake prediction over a long period in which we have sufficient anomalies and earthquakes, let us define m, n, and μ as follows;

m: The number of cases for which an earthquake occurs after observing an anomaly and issuing a prediction.

n: The number of cases for which no earthquake occurs in spite of observing an anomaly and issuing a prediction.

μ: The number of cases for which an earthquake occurs without any premonitory anomaly.

In that case the total number M of earthquakes is given as

$$M = m + \mu \tag{5.6}$$

The number of issuance of prediction becomes

$$F = m + n \tag{5.7}$$

The probabilities of success and prediction are therefore obtained as

$$p_1 = m/F, \qquad p_2 = m/M \tag{5.8}$$

The total amount of earthquake damage during the period concerned becomes ML when no prediction is made. On the other hand, when predictions are made, the sum of the amount of earthquake damage and expenses for prediction work is given as $Ef + ml + \mu L + C$. The quantity E that expresses the effect of earthquake prediction is defined by

$$E = [ML - (Ef + mL + \mu L + C)]/ML \tag{5.9}$$

It is clear that $E > 0$ is necessary. If $E < 0$, predictions are harmful.

Equation (5.9) can be rewritten as

$$E = p_2 \left(1 - \frac{1}{L} - \frac{1}{p_1} \frac{f}{L} \right) - c \tag{5.10}$$

where

$$c = C/(ML) \tag{5.11}$$

Should p_1 and p_2 tend to 1, E would tend to E_0 which is given as

$$E_0 = 1 - \frac{1+f}{L} - c \tag{5.12}$$

On assuming an ideal case for which $p_1 = p_2 = 1$, it becomes known that the earthquake damage can be lessened by $E_0 \times 100\%$ by issuing earthquake predictions.

The above theory as developed by UTSU (1977) is interesting. It is difficult, however, to apply the theory to an actual problem because the estimate of various parameters involved is hard to perform. In spite of such difficulties, the author dares to estimate the effect of earthquake prediction by making use of data in Shizuoka Prefecture under the menace of the feared Tokai earthquake.

The Shizuoka Prefectural Office publicized a report on the estimate of anticipated damage due to the expected Tokai earthquake in November, 1978. According to the report, some 418,000 families will be afflicted by the earthquake disaster in the prefecture when no prediction is made. The figure would decrease to 221,000, should a prediction be made because of proper reaction of inhabitants. Let us assume that each afflicted family lost a sum of 10 million yen. In that case, L and l are estimated as

$$L = 4.18 \times 10^{11} \text{ yen}, \qquad l = 2.21 \times 10^{11} \text{ yen}$$

As only one prediction is made, the parameters take on the following values.

$$M = 1, \quad m = 1, \quad \mu = 0, \quad F = 1, \quad p_1 = 1, \quad p_2 = 1$$

Both the expenditures for the countermeasures against earthquake disaster and the observations for prediction are tentatively assumed to amount to 10 billion yen, so that we have

$$f = C = 10^{10} \text{ yen}$$

With all these figures for various parameters, E is estimated as

$$E = 1 - \frac{2.21 \times 10^{11} + 0.1 \times 10^{11}}{4.18 \times 10^{11}} - \frac{0.1 \times 10^{11}}{4.18 \times 10^{11}} = 0.42$$

The above estimate does not include public facilities, private enterprises and the like. Should these things be included, the estimated value of E would increase. It is no easy matter to calculate E with p_1 and p_2 evaluated on the sound basis. The above estimate of E is made just for seeing a very rough value of E.

5.3.2 Probability of success due to observation of a number of precursory phenomena

Let us imagine that there are two earthquake prediction elements, A and B. Let us further suppose that the numbers of earthquakes, that occurred after observing anomalies for A and B, are m_A and m_B. When A and B are independent each other and the total number of earthquakes is M, the expectancy of the number of cases, for which an earthquake occurs in association with anomalies both for A and B, is estimated as $m_A m_B / M$.

Let us next denote the numbers of cases, for which we have no earthquake in spite of anomalies for A and B, by n_A and n_B. UTSU (1977) presumed that the expectancy of the number of cases, for which no earthquake occurs in spite of the fact that anomalies are observed both for A and B, is proportional to n_A and n_B. When the proportional constant is taken as $1/N$, the expectancy is given as $n_A n_B/N$.

In that case the probability of an earthquake to occur when anomalies are found both for A and B is estimated as

$$p(A, B) = \frac{m_A m_B/M}{m_A m_B/M + n_A n_B/N} \tag{5.13}$$

This is nothing but the probability of success or p_1 (A, B). But the suffix 1 is omitted hereafter.

Equation (5.13) can be rewritten as

$$p(A, B) = \frac{1}{1 + R\left(\frac{1}{P_A} - 1\right)\left(\frac{1}{P_B} - 1\right)} \tag{5.14}$$

where p_A and p_B are the probability of success when an anomaly appears for A and B separately. R is defined by

$$R = M/N \tag{5.15}$$

According to Utsu, when a prediction is issued randomly, the probability of success is given by

$$p_0 = \frac{M}{M + N} \tag{5.16}$$

It is clear that the number of cases for which no earthquake occurs when anomalies are observed both for A and B is $n_A n_B/N$. N is nothing but the proportional constant already introduced.

Eliminating M/N from Eqs. (5.15) and (5.16), we obtain

$$R = \frac{1}{\frac{1}{p_0} - 1} \tag{5.17}$$

Putting Eq. (5.17) into Eq. (5.14), we arrive at

$$p(A, B) = \frac{1}{1 + \left(\frac{1}{p_A} - 1\right)\left(\frac{1}{p_B} - 1\right)\bigg/\left(\frac{1}{p_0} - 1\right)} \tag{5.18}$$

If we observe anomalies for earthquake prediction elements A,

B, C,···, S, the synthesized probability of success becomes

$$p(A, B, C, \cdots S) = \frac{1}{1 + (1/p_A - 1)(1/p_B - 1) \cdots (1/p_S - 1)/(1/p_0 - 1)^{s-1}}$$

(5.19)

The situation is much the same as Eq. (5.3). When anomalies appear for a number of elements, the synthesized probability becomes higher and higher accordingly as the number of anomaly increases.

UTSU (1979) attempted to estimate the probability of success taking the Izu-Oshima Kinkai earthquake ($M = 7.0$, 1978) as an example. The following three phenomena are assumed to provide earthquake precursors.

A) The anomalous land uplift (see Fig. 4.1) found on the Izu Peninsula is assumed to be a kind of precursor. The probability of a $M \geq 6.5$ earthquake occurring within 5 years' time since the outbreak of the anomaly is assumed as 1/3.

B) The seismic activity in the morning of January 14, 1978 is regarded as foreshocks with a probability of 1/10. According to Utsu, 19 of the 26 foreshock activities in Japan had the precursor time equal to or smaller than 3 days, and the difference in magnitude between the main shock and the next large shock exceeds 1.6 for 10 of them. If we assume that the magnitude of main shock amounts to 6.5, the difference in magnitude between the main shock and the next large shock becomes 1.6 because the maximum magnitude for foreshocks in the morning of January 14 was 4.9. It is therefore estimated that a $M \geq 6.5$ earthquake will occur within 3 days' time since the outbreak of foreshock activity with a probability amounting to $(1/10) \cdot (19/26) \cdot (10/20) \doteqdot 1/35$.

C) The anomalies of rock volume as observed by a borehole volume strainmeter buried at Irozaki, the anomalies of radon concentration and underground water level observed at Nakaizu are treated as a group of precursors. It is not known whether they are independent of each other. On the basis of these anomalies, it is estimated that a $M \geq 6.5$ earthquake will occur with a probability of 1/10 within one month's time since the middle of December last year.

In order to evaluate p_0, let us suppose the following two cases.

Case I: It is assumed that a $M \geq 6.5$ earthquake occurs once in

30 years in an area including Odawara, Izu-Oshima Island and Izu Peninsula.

Case II: An area, which is more limited than the above, is under consideration, and so it is assumed that a $M \geq 6.5$ earthquake occurs once in 100 years.

Let us assume that a prediction is made on the basis of the data cited in the above at about 11 o'clock on January 14. If it is assumed that a $M \geq 6.5$ earthquake should occur, p_A, p_B, p_C, p_0, and $p(A, B, C)$ are estimated as given in Table 5.1. In the table we observe that the so-called synthesized probability $p(A, B, C)$ becomes fairly high even if p_A, p_B, and p_C representing the estimate due to an individual anomaly are not quite high. The point that the synthesized probability becomes very high by combining a number of mutually-independent prediction elements is interesting. Even so, however, the estimated values of $p(A, B, C)$ as presented in Table 5.1 seem so high that they are much higher than those based on intuition. A more accurate estimate of p_A, p_B, p_C, and, consequently, $p(A, B, C)$ will be effected by more intensified observation system in the future.

TABLE 5.1. p_A, p_B, p_C, p_0 and $p(A, B, C)$ for three values of τ as estimated for the Izu-Oshima Kinkai earthquake (UTSU, 1979).

		3 hours	1 day	3 days
	p_A	0.0000228	0.000183	0.000548
	p_B	0.00119	0.00952	0.0286
	p_C	0.000417	0.00333	0.01
p_0	Case I	0.0000114	0.0000913	0.000274
	Case II	0.00000342	0.0000274	0.0000821
$p(A, B, C)$	Case I	0.080	0.41	0.68
	Case II	0.49	0.89	0.96

5.4 The Four Stages of Earthquake Prediction—Algorithm of Prediction

Summarizing what the author stated in Chapters 2 to 4 and the foregoing sections in this chapter, it appears that the approach basing on the following four stages would be the most appropriate approach to a quantitative prediction so far as the author's view is concerned. One may feel that the following discussion seems to rely on an incomplete quantification of probability estimation. Even the author feels that such a view is right. But the author's approach would become realistic when more data are accumulated and a concrete model of focal process is brought to light. It is the author's belief that even the strategy of CCEP would become something like what follows when reliable quantifications of estimate of earthquake risk is developed.

5.4.1 Statistical prediction—preliminary stage

As stated in Chapter 2, probability can be assessed in some cases about occurrence of a large earthquake on the basis of migration of epicenters, seismic gaps and so on. It should be borne in mind, however, that a statistical view is applicable only to the average state of phenomena that occurred in the past. It is quite possible that a particular instance that will occur next time may deviate very largely from the averaged state. The author believes, however, that occurrence of large earthquakes at subduction zones, which is caused by the subdution—rupture—rebound process, as discussed in Chapter 2, may be a good target of statistics, should a fairly long historical record of earthquake occurrence be available. In such a case, the probability of large earthquake recurring at a certain area of a subduction zone can sometimes be estimated.

5.4.2 Long-term prediction—strain accumulation stage

If it is assumed that the prime cause of large earthquakes is due to compression and down-pulling of a land plate by subduction of a sea plate, it is quite understandable that we can say something about occurrence of a large earthquake provided that the accumulation of crustal strain can be monitored. As mentioned in Chapter 3, the knowledge on the ultimate strain to rupture the earth's crust is also indispensable for discussing the possible occurrence of a large earthquake. In a number of areas in Japan and California, where geodetic surveys have been frequently carried out, it is actually possible to estimate the probability of a large earthquake recurring at a certain area.

Should the precise geodetic surveys as proposed by GSI be completed and repeated within a reasonable period of time, it would be possible to estimate the probabilities of an occurrence of inland earthquake having a magnitude of 7 or so, to some extent throughout Japan.

5.4.3   Medium- and short-term prediction—stage for appearance of precursors of the first kind

When we conduct a highly-dense observation over an area where the probability of an earthquake occurrence is estimated as fairly high by either analysis of historical records or monitoring of crustal strain as stated in the last two subsections, there is good reason to believe that precursory effects of some sort can be detected sooner or later. They are firstly precursors of the first kind as classified in Section 4.11.

When the spatial extent of such a precursor can be identified, empirical equations (5.1) or (5.2) may facilitate a possible estimate of earthquake magnitude M. Once M is estimated, another empirical equation (4.3) enables us to estimate the precursor time of the supposed main shock. As these equations utilized here are empirical ones, the estimate of magnitude as well as occurrence time of the main shock can be made only probabilistically.

It is in theory possible to obtain a fairly accurate probability of earthquake occurrence by calculating a synthesized probability on the basis of precursors of various disciplines as discussed in Section 5.2. It turns out, however, that an estimate of such a synthesized probability cannot easily be performed because of lack of data.

5.4.4   Imminent prediction—stage for appearance of precursors of the second and third kinds

When the time of earthquake occurrence, which is deduced by the appearance of precursors of the first kind, approaches, it would be highly probable that we observe precursors of the third and second kinds. In that case a prediction may be issued on a scientifically firm basis. It should be emphasized that precursory signals are apt to be contaminated by natural and man-made noise. Hence, a prediction should be made on the basis of a number of precursory effects which are independent from one another.

5.5 Chinese Strategy of Earthquake Prediction

On the basis of what the author could hear and see when he and his missions visited China in 1978 and 1980, it will be attempted in this section to review the Chinese strategy of earthquake prediction. Reports by OIKE (1978a, b, c, 1979) and SUZUKI (1978a, b, c, d) are of great help for writing the manuscript of this section.

The Chinese strategy of earthquake prediction is more or less the same as that discussed in the last section. Earthquake prediction information is classified into four stages, i.e. long-, medium-, short-term and imminent predictions. A long-term prediction means the designation of a certain area, in which the probability of occurrence of a large earthquake is high, to an area of accentuated observation. Such a designation is usually made by the National Conference on Earthquake Prediction. For instance, a fairly wide area such as North China and Bohai area is usually designated such an area. A designation is sometimes made by a conference for particular regions such as the three provinces in North-east China.

The basis of a long-term prediction is seismotectonic examination of active faults, analysis of seismic activity including migration of epicenter, seismic gap and the like, and instrumental observations at seismological observatories.

In China, a prediction over a period of several months to 1–2 years prior to the main shock is called the medium-term prediction. Meanwhile a short-term prediction covers a period of several weeks to several months prior to the main shock. It is said that a medium-term precursor occurs somewhat slowly while a short-term one occurs much more rapidly. Almost all phenomena discussed in Chapter 4 are regarded as precursors. Definition of an anomaly is in many cases not clear. It appears that there is no well-established rule for defining an anomaly. Whenever something anomalous is observed, they generally take it as a precursor in China. Such an attitude is completely different from that in Japan. When an anomalous phenomenon is observed in Japan, scientists first question whether it is a false signal. Table 5.2 is a list of long-, medium-, and short-term predictions in China as summarized by OIKE (1978c).

When the precursor time is supposed to be several minutes, hours

TABLE 5.2. Examples of long-term, medium-term, and short-term prediction (OIKE, 1978c).

No.	Date	Conference	Prediction	Basis	Effect
1	1970	National Conference	Designated of regions, Liaoning, Hebei Province	Increasing regional seismicity, migration	Arrangement of seismological brigades and stations
2	June 1974	Regional, N. China and Hebei	Northern part of Hebei, 1–2 years, M 5–6	Leveling, geomagnetism, seismicity	Forecast issued to the public, acceleration of observation work
3	Nov. 1974	Three Prov. in NE China	Yingkou-Dalian county, early date, disastrous event	Short-span leveling geomagnetism, increasing micro-earthquakes	Prediction network
4	Dec. 1974	Liaoning Province	South-Liaoning, early date, M 4–5	Animal behavior, ground tilts	After M 4.8, 22 Dec. emergency measures
5	Jan. 1975	National Conference	Yingkou-Jin Xian, and Dando, the first half of 1975, M 5.5–6	Animals, wells, short-span leveling, seismicity gap	Patrol observation stations

TABLE 5.2 (continued)

No.	Date	Conference	Prediction	Basis	Effect
6	Jan. 1975	National Conference	West Yunnan Prov. 1–2 years, $M7$	Seismicity	Organizing of observation system, training of the people
7	Nov. 1975	Yunnan Province	Bijiang, Lushui, Longling, Ruili, $M6$	Radon, tilt, stress, seismicity	
8	Nov. 1975	Sichuan Province	Songpan-Maowen, the first half of 1976, $M>6$	Past few years data	Authorized by the National Conference of January 1976, establishment of network
9	Feb. 1976	Yunnan Province	Existence of abnormalities	Radon, tilt, stress, seismicity	A directive to 7 prefectures, acceleration of preventive work
10	13 May 1976	Yunnan Province	Central part of Xiao Jiang fault, the last half of May, $M5$–5.5 Western Yunnan, $M7$	Synthesized data from 26 local units	A directive to accelerate work released 15 May

TABLE 5.2 (continued)

No.	Date	Conference	Prediction	Basis	Effect
11	June 1976	National Conference	Peking, Tianjin Tangshan, $M > 6$	Continuous abnormalities after the Haicheng earthquake	Investigation party to the Tangshan region
12	14 June 1976	Sichuan Province	From the central part of Longmen Shan fault to Kangding, 1–2 months, $M\,6$	29 May event in the Yunnan region	
13	22 June	Urgent N-S Earthquake-Belt Conference	The central part of Longmen Shan fault, till the end of Aug., $M > 6$		Increase of observation points by the masses, temporary stations, sending experts to the region
14	2 Aug. 1976	Sichuan Province	Maowen-Beichuan and surroundings or Kanding, 13–17 and 22 August, $M\,6$–7		Information to the local offices in the related regions
15	15 Sept. 1976	Sichuan Province	Yanyuan-Xichang, Nov. 1976–Jan. 1977, $M\,7$		

TABLE 5.3. Examples of imminent predictions (OIKE, 1978c).

No.	Main shock	Place	Prediction
1	5 Dec. 1967 night	Shulu, Hebei	An observer predicted on the basis of an abnormality of ground-water that day
2	18 July 1969 $M\,7.4$	Hebei	A breeder of the Tianjin zoo reported to the Tianjin Earthquake Office on unusual animal behavior
3	8 Nov. 1970 $M\,5.5$	Barkam, Sichuan	Mulang Commune, felt foreshocks and unusual animal behavior; saving lives
4	March 1971 $M\,6.3$	Xinjiang-Uygur	The imminent stage; preventive measures taken
5	25 Apr. 1971 $M\,3.9$	Shizuishan, Ningxia	The Mine Bureau, abrupt changes of ground-water level, "$M\,3$–4, in five days" on 22 April
6	11 June 1971 $M\,4.9$	Ditto	Wuzhong County, foreshocks and unusual animal behavior 10 minutes before event
7	5 Aug. 1971 $M\,3.4$	Xingtang, Hebei	Hebei Seismological Bureau, from the formula for radon, "6–8 Aug. in the Shanxi region, $M\,4.5$" on 28 July
8	16 Aug. 1971 $M\,5.8$	Mabian, Sichuan	Preventive measures taken
9	26 Aug. 1971 morning $M\,4$	Shulu, Hebei	Ground-water, reported "event in three days" on 24 August
10	23 Jan. 1972 10h $M\,5.5$	Honghe, Yunnan	Earthquake Office of Honghe Auton. Pref., data from stations and from indigenous station at Jianshui, "Western part of the Pref. $M\,4.5$ around 20 Jan." issued to the surrounding regions on 16 January

TABLE 5.3 (continued)

No.	Main shock	Place	Prediction
11	3 Feb. 1972 15h22m, 24m M 4.8, M 4.2	Ditto	Decreasing swarm, earth currents, fish, temperature, "M 4.5, in 24 hours", at 18 hours before event, reinforced, and preventive measures taken
12	7 Sept. 1972 M 4.8	Xingjiawan, Hebei	Successfully predicted
13	27 Sept. 1972 8h08m M 5.8	Kangding, Sichuan	Local people, from earth-currents and animals, "M 5.5, near place" in the afternoon of 26 September
14	30 Sept. 1972 0h M 5.8, 5.5	Ditto	Continuous abnormalities after the above event,"Big event again"
15	12 Oct. 1972 M 5.2	West Shahe	Successfully predicted
16	22 March 1973 M 5.5	Simao, Yunnan	Yuxi Mail and Telegraph Office, tilt, earth-currents and ground-water, "M 4–5, Eshan, Yuan Jiang, Simao region, in a few days", issued to counties by telephone, on 20 March, preventive measures taken
17	24 March 1973 M 5.7	Garze, Sichuan	Preventive measures taken with the imminent warning
18	2 June 1973 M 5.3	Tengchong, Yunnan	A staff member of the meteorologica observatory, from the relation between drought and shocks, "M 5.5–6, June, Tengchong", at the end of May, preventive measures taken
19	16 Aug. 1973 M 6.2,	Pu'er, Yunnan	Pu'er Earthquake Office, abrupt changes of earth-currents, "Possibility of big event", telephoned to communes, preventive measures taken
20	9 Sept. 1973 M 5.7	Garze, Sichuan	Preventive measures taken with the imminent warning

TABLE 5.3 (continued)

No.	Main shock	Place	Prediction
21	11 May 1974 M 7.1	Chaotong, Yunnan	A commune, animals, rumbling and flash, sheltered and guarded from the evening of 10 May
22	6 June 1974 M 4.9	Ningjin, Hebei	Hebei Seismological Bureau, radon data at Hongshan and Xingtai, "$M>4$, 31 May–6 June, aftershock region of Xingtai", on 30 May
23	18 July 1974 M 4.3	Hebei	Ningjin Seismological Observatory, abnormality in ground-water, predicted on 14 July
24	4 Feb. 1975 19h36m M 7.3	Haicheng, Liaoning	Liaoning Revolutionary Committee, by the synthesized conclusion, "Haicheng, Yingkou region, a big event", a directive of prevention issued to counties at 10h, 4 Feb.
25	29 May 1976 M 7.5	Longling, Yunnan	Yunnan Seismological Bureau, the synthesized data, "The end of May– the beginning of June, a destructive event", on 28 May, the imminent warnings were 20 minutes before the main shock
26	28 July 1976 M 7.8	Tangshan, Hebei	An engineer of the 129 express train seeing a flash, braked up in front of a bridge 1 minute before the main shock
27	16 Aug. 1976 M 7.2	Songpan, Sichuan	Sichuan Seismological Bureau, abrupt increase of abnormalities, the imminent warning on 12 August, taking a documentary picture of abnormal phenomena into account
28	7 Nov. 1976 M 6.9	Sichuan, Yunnan, Border	Sichuan and Yunnan Seismological Bureaus, "M 6–6.5, around 7 Nov. Ninglang, Muli and Yanyuan region," on 3 Nov., preventive measures taken
29	13 Dec. 1976 M 6.8	Ditto	Reappearance of abnormal phenomena "Big event again", issued by posters to the people who had been taking shelter, on 6 Dec.

or days, the prediction is called an imminent prediction which is related to a warning and evacuation. Earthquake precursors for this class of prediction are more or less the same as those for medium- and short-term predictions. However, these precursors turn up quite suddenly and the number of reports increases enormously. It is also seen that precursors tend to concentrate towards the epicentral area of coming earthquake as time goes on. The author understands that there is no established formula which correlates a precursory phenomenon to an earthquake occurrence at a certain date.

As the time of earthquake occurrence approaches, Chinese seismologists begin to feel a highly-strained situation because of increase in number of reported anomalies, magnitude of anomalous changes in instrumental and macroscopic phenomena, and rate of concentration of these precursory effects. Finally, they decide to present the information about imminent prediction to local and regional administrators. According to Chinese colleagues, they believe that such a decision is a kind of synthesized judgment.

The situation is much the same as the judgment by a war council because decision must be made on the basis of insufficient data, so that subjective elements due to perception and experience of persons in charge may play some role in actual judgment. In spite of such defects, the point that Chinese colleagues succeeded in many imminent predictions should be appreciated. Table 5.3 is the list of imminent predictions in China as summarized by OIKE (1978c).

5.6 U.S. Strategy of Earthquake Prediction

The seismic activity being lower in the U.S.A. than that in Japan and China, no concrete strategy of earthquake prediction has as yet been put forward so far as the author is aware. It appears that USGS is working mostly on observing earthquake precursors along with clarifying their space-time characteristics. At the moment, no ambitious strategy of earthquake prediction has been developed by American colleagues.

5.7 Soviet Strategy of Earthquake Prediction

Soviet colleagues seem to put much stress on earthquake pre-diction based on multi-element observation. According to ASIMOV *et al.* (1979), for example, a magnitude 7.0 earthquake that occured in

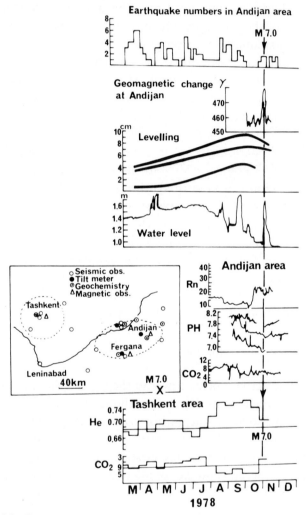

FIG. 5.5. Precursory changes in various elements for earthquake prediction prior to the Pamir Plateau earthquake ($M = 7.0$, Nov. 1, 1978). Units of earthquake numbers, CO_2 and He are not known. (ASIMOV *et al.*, 1979).

Pamir Plateau on November 1, 1978 was predicted on the basis of the changes in seismicity, geomagnetic field, underground water level, radon content and other geochemical elements in the Tashkent–Fergana–Andijan area in the Uzbek Republic, Middle Asia, the changes of these disciplines being reproduced in Fig. 5.5.

It is said that the scientists at the Uzbek Academy of Sciences issued a warning that an earthquake would occur to the south of Andijan within 6 hours' time in the evening on October 30 although no exact epicenter could be located because of the lack of observation in the southern Pamir. Looking at Fig. 5.5, however, the author is not quite convinced that the data are good enough for such a prediction. The author hopes that Soviet colleagues will publish papers in which the processes of prediction are described in great detail.

It is rather surprising that Soveit colleagues admit that they can forecast the occurrence time of a strong earthquake. But they cannot pinpoint the epicenter. This is probably due to the fact that many earthquakes occur outside the observation networks.

5.8 Comments on the Existing Strategies of Earthquake Prediction

It is really surprising that earthquake prediction, that used to be the business of astrologers and fortune-tellers some 20 years ago, has developed to such an extent that its strategy can be argued. But, the strategies proposed are very much incomplete at the present stage.

ISHIBASHI (1978) pointed out that the emprical relation between logarithmic precursor time $\log_{10} T$ and earthquake magnitude M as expressed by Eq. (4.3) is subjected to a large change in precursor time when the coefficients involved are slightly improved by introducing new data. This is very much so for an earthquake of large magnitude. Rikitake's previous formula (RIKITAKE, 1975c) leads to a precursor time amounting to 49 years for a $M=8$ earthquake, while Eq. (4.3), which is recently improved by RIKITAKE (1979d), indicates 17 years. Such a large difference in the estimate of precursor time is fatal for practical prediction.

Ishibashi further doubts the applicability of Eqs. (5.1) and (5.2) as proposed by DAMBARA (1966, 1981) for the relationship between the area over which precursory effects are prevailing and earthquake magnitude. According to Ishibashi, therefore, the strategy of earthquake prediction as developed by the present author and summarized in Section 5.4 is only an idealistic approach that cannot be applied to an actual prediction.

In general what Ishibashi said is right. The author admits of drastic idealization which was necessary for quantifying the approach. The author thinks that the coefficients of Eq. (4.3) may differ from region to region depending upon the seismogenic characteristics. It is regrettable that available data are so few that no statistics for different regions can be separately conducted, so that Eq. (4.3) reflects only the general tendency averaged for precursors appeared in various parts of the world. It is possible, therefore, that the precursor time versus magnitude relationship of a particular earthquake deviates considerably from the mean tendency as expressed by Eq. (4.3) and the straight line in Fig. 4.54. Very few data for earthquakes having a magnitude of 8 or thereabout are included in deducing Eq. (4.3), so that Eq. (4.3) may be good only for a magnitude range 5 to 6 for which much data exists. It is urgently required for future earthquake prediction research to work out precursor time versus magnitude relationships such as Eq. (4.3) for various seismotectonic units.

In conclusion we have to admit that no established strategy of earthquake prediction is available at present. In an actual prediction, therefore, we have to rely partly on the poorly established relationship such as Eq. (4.3) along with historical documents, migration of epicenters, concept of seismic gap, concentration of macroscopic anomaly and so on. As has been emphasized by Chinese colleagues, this is a kind of synthesized judgment. In order to accomplish scientifically supported prediction, every element involved in the synthesized judgment should be quantified as much as possible. Such quantification may only be achieved by promoting the earthquake prediction program to a greater extent.

CHAPTER 6

ORGANIZATION AND SYSTEM OF EARTHQUAKE PREDICTION

The earthquake prediction program in Japan is underway under the cooperation of existing governmental institutions such as Japan Meteorological Agency (JMA), Geographical Survey Instisute (GSI), Earthquake Research Institute (ERI) of the University of Tokyo, Geological Survey of Japan, National Research Center for Disaster Prevention, and other national universities. These organizations, are concerned with a number of studies other than earthquake prediction, and it is natural that rather smaller percentages of labor, manpower, and budget are expended on earthquake prediction in most institutions.

Even though CCEP was set up in 1969, its function is only for "coordination" as its name indicates. Under the circumstances, the system of earthquake prediction in Japan involves various difficulties because no organization is entitled to control everything related to earthquake prediction.

This chapter aims at describing the systems of earthquake prediction in relevant nations along with the difficulties involved. Also included is a review of the desirable system of earthquake prediction in Japan based on the author's personal point of view.

6.1   System of Earthquake Prediction in Japan

6.1.1   Geodetic Council

Japanese programs related to geophysical observations are administratively coordinated and adjusted by the Geodetic Council attached to the Ministry of Education. The Japanese participation to the International Geophysical Year, International Upper Mantle Project, International Geodynamics Project and the like materialized through discussion and authorization by the Council. It is customary for the Council to raise a proposal to the prime minister and other ministers concerned about the budget and necessary arrangements required for carrying out a nation-wide project. The Japanese program on earthquake prediction has also been promoted in a manner similar to those mentioned above.

Since earthquake prediction work has been steadily developing in recent years reaching a fairly large annual budget and, at the same time, the academic characteristics of the work gradually replaced by more practical approaches to actual prediction, one may think that the coordination of the work had better be handled by a little more powerful committee which is concerned with technology for practical use, rather than scholarly achievement. However, the effort hitherto made by the Geodetic Council and the Ministry of Education through which the Japanese earthquake prediction program has reached the present level of development should be highly appreciated.

6.1.2 Subcommittee on Earthquake Prediction

A Subcommittee on Earthquake Prediction has been set up in the National Committee for Geodesy and Geophysics, Scince Council of Japan since 1965 (see Table 1.2). Unlike the Geodetic Council, discussion regarding earthquake prediction is made freely on the basis of a purely academic standpoint at the subcommittee. At the early stage of the earthquake prediction program, the Subcommittee played an important role in preparing the draft plan of the program. It was customary for the Geodectic Council to adjust the planning presented by the Subcommittee under administrative consideration.

As the earthquake prediction program became a sort of routine enterprise, funded by a large-scale budget reflecting public demands, the Subcommittee's effort is now concentrating on debating what the earthquake prediction program should be, especially in the future, from an academic viewpoint. This debate over preparing the second blueprint of earthquake prediction program or the long-range program on earthquake prediction will be stressed at the Subcommittee. Another important role of the Subcomittee is the international liaison of earthquake prediction. The Subcommittee also sponsors national symposia on earthquake prediction held every 2 to 3 years.

6.1.3   Coordinating Committee for Earthquake Prediction
1)   Establishment

CCEP was eatablished in GSI in 1969 as has already been stated in RIKITAKE (1976a) and briefly mentioned in Subsection 1.1.1 (see Table 1.2).

Item No. 6 of separate sheet No. 2 of the proposal for the second 5-year program on earthquake prediction by the Geodetic Council on July 16, 1968 refers to the overall system for promoting the program. It was mentioned in the item that close cooperation between the organizations involved, efficient analysis of observed data and synthesized judgment are required for promoting smoothly and efficiently the program arriving at quick and composite conclusions. It was further stressed in Item No. 6 that a central agency for promoting the program should be formed by setting up a coordinating committee to handle the interchange of data and information between the organizations involved in the program and, at the same time, be involved in the overall evaluation and judgment of information available.

On the basis of this proposal of the Geodetic Council, it was finally decided in April, 1969 to establish CCEP as an advisory organ to the director of GSI. It was a matter of much debate to which organization CCEP should be attached. As only long-term prediction was thought feasible at that time, it was decided to attach the committee to GSI because geodetic survey, which is handled by GSI, is the most powerful means for achieving long-term prediction.

No particular legislation was made for establishing CCEP. Officially speaking, therefore, the committee is only a personal advisory organ to GSI's Director. No sufficient budgetary arrangement has been made, so that some of the members of the committee suffer from a shortage of travel expenses for attending committee meetings held several times each year. The members amounting to 30 in total number are all university professors and officials at governmental institutions, so that they are all working for the committee on a part-time basis. It is therefore difficult to say that CCEP is in a position to take the full responsibility for earthquake prediction in Japan.

2)   Particulars, construction and management

The flow of data to CCEP from respective organizations working on earthquake prediction is shown in the block-diagram in Fig. 6.1.

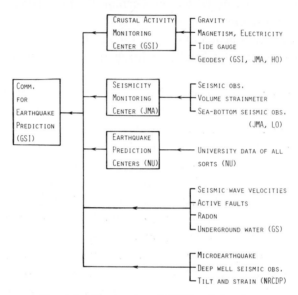

FIG. 6.1.   Construction of CCEP and data flow.

In the early stage of CCEP, most data observed were sent to CCEP through the three centers, i.e. Crustal Activity Monitoring Center of GSI, Seismicity Monitoring Center of JMA and Earthquake Prediction Observation Center of ERI. Data concerning gravity, magntism, electricity, tide and geodesy mostly taken by GSI and partly by JMA and the Hydrographic Office (HO) are processed at the Crustal Activity Monitoring Center. Seismic data taken by JMA and Latitude Observatory (LO) as well as of the data taken by JMA are processed and sent to CCEP through the Seismicity Monitoring Center. It was originally planned to process data taken by universities at a number of provincial sub-centers and to send those data to CCEP through the Earthquake Prediction Observation Center. But such sub-centers evolved enough in recent years, so that they may now be called Earthquake Prediction Centers which can send data direct to CCEP. The center attached to ERI has become especially well-equipped with a high-speed computer and other facilities and is now named the Earthquake Prediction Data Center which is supposed to coordinate and summarize all the data taken by university proups.

The Geological Survey of Japan (GS) and the National Research Center for Disaster Prevention (NRCDP) have also been actively participating to the earthquake prediction program. As can be seen in Fig. 6.1, the data of various disciplines taken by them go direct to CCEP.

Figure 6.1 also brings the construction of CCEP to light very clearly. All the arrows on the diagram are pointing to CCEP. This means that CCEP gets data from the organizations involved, but CCEP is not in a position to give an order to them. This is one of the weak points of CCEP. Even if CCEP concludes that a particular observation should be urgently made over a certain area, it has no power to give an order to relevant organizations about the proposed work. The president of CCEP may ask the organizations concerned to promote such an observation, but he has no funds to make the work possible.

Although it is becoming the national consensus that CCEP functions as the headquarters of earthquake prediction, it has no authorized power as stated in the above. Strictly speaking, CCEP is not entitled to issue an earthquake prediction information. It has become customary that CCEP's president holds a press conference after its meetings. He may say, for example, that an anomalous uplift is found at a certain area, but nothing more officially. Of course, pressmen would write an article that, should the uplift be connected to an earthquake, the magnitude would be so and so and the time of occurrence would be expected within so and so years. It should be seriously noted, however, that, legally speaking, even the president of CCEP has no right and obligation to make an earthquake prediction.

In spite of such weakness of CCEP administration, the author believes that cooperation between organizations involved in earthquake prediction work has so far been excellent, so that an enormous amount of data pertinent to earthquake prediction has been accumulating in Japan. The raw data are regularly published in the Report of the Coordinating Committee for Earthquake Prediction in Japanese.

CCEP decided the outline of its own management at the 34th meeting on August 23, 1976 as follows;

i) The committee consists of members less than 30 in number. It is possible to appoint temporary members when necessary.

ii) Regular and temporary members should be men of learning and experience or officials of relevant government institutions. They are commissioned by GSI's Director.

iii) A working sub-committee may be set up when a particular matter is to be examined. A working sub-committee shall consist of regular and temporary members.

iv) The term of membership be two years. When a substitute member is commissioned because of vacancy, the term of membership shall be the remaining period for the predecessor.

v) The president of the committee shall be elected by mutual vote. When the president is not available, a committee member, who has been nominated by the president in advance, shall take the chair. The chairman of working sub-committee shall be nominated by the president.

vi) The committee may have several honorary consultant members. They are commissioned by GSI's Director on the recommendation of the president.

vii) The president calls a committee meeting when necessary. The working sub-committee shall be called by the chairman of the sub-committee.

viii) Report and announcement related to synthesized judgment by CCEP shall be published by GSI. However, information about actual occurrence of large, medium and small earthquakes shall be announced by JMA.

xi) Other necessary matters in relation to the management of the committee shall be decided by the president after discussion at a committee meeting.

x) Administrative business of the committee is handled by GSI.

6.1.4 Headquarters for Earthquake Prediction Promotion
As can be seen in Table 1.2, a body that coordinates earthquake

prediction administration among various ministries involved was established in 1973. The body was strengthened and changed into a committee headed by the Vice-minister of Science and Technology in 1974.

When social demands for earthquake prediction became very strong in association with the increasing fear of possible occurrence of a great earthquake in the Tokai area, the Japanese government decided to set up the Headquarters for Earthquake Prediction Promotion on the basis of a decision at a cabinet meeting on October 29, 1976. The former body for administrative coordination of earthquake prediction was replaced by the Headquarters.

The decision of cabinet meeting reads as follows;

i)   In order to promote administrative businesses between ministries and agencies involved and to push overall and intentional policies in relation to earthquake prediction, the Headquarters for Earthquake Prediction Promotion shall be attached to the cabinet.

ii)   The Headquarters shall discuss and promote the following items:

a)   Promotion of researches on practical use of earthquake prediction

b)   Systems necessary for promoting earthquake prediction such as concentration of data and the like

c)   Items which should be put into action urgently and intensively in regard to promotion of earthquake prediction

d)   Materialization of the proposal of the Geodetic Council for earthquake prediction

e)   Promotion of researches on communication of earthquake prediction information and on technology of disaster prevention which may become urgently necessary because of development of earthquake prediction

f)   Other items related to promotion of earthquake prediction

iii)   The construction of the Headquarters is as follows.

The General Director may add members when necessary.

General Director: Minister of Science and Technology

Members: Vice-Director of Cabinet Secretariat

Vice-Minister of Science and Technology

Vice-Minister of National Land Agency

Vice-Minister of Education
Vice-Minister of International Trade and
   Industry
Vice-Minister of Transportation
Vice-Minister of Construction
Vice-Minister of Autonomy

When the General Director is not available, a member who has been nominated beforehand by the General Director shall take the chair.

iv)   A secretary of the Headquarters be nominated by the General Direcor. The Secretary must be an offical of governmental organizations involved.

v)   The General Director may ask men of learning and experience in the field of earthquake prediction to attend a meeting of the Headquarters in order that the Headquarters members may listen to their comments.

vi)   Administrative businesses of the Headquarters are handled by governmental organizations involved, but the general control will be made by the Science and Technology Agency.

vii)   Items necessary for managing the Headquarters other than the last six items shall be decided by the General Director.

The author being an earth-scientist, he does not quite understand how the earthquake prediction work would be benefited by creating an organization such as the Headquarters concerned. It is certainly a good thing to establish such a Headquarters provided that it backs up the promotion of earthquake prediction by making efforts toward getting an adequate budget and manpower through administrative coordination.

In Fig. 6.2 is shown the block-diagram of the complicated system of earthquake prediction now underway among the governmental institutions in Japan. The Headquarters is an organ that plans to make administrative coordination between such a complex system. But it is the author's understanding that the coordination by the Headquarters is only for administrative matters. It is not qualified for any scientific control.

The author believes that readers would certainly be in a state of confusion because a number of committees, councils and the like are

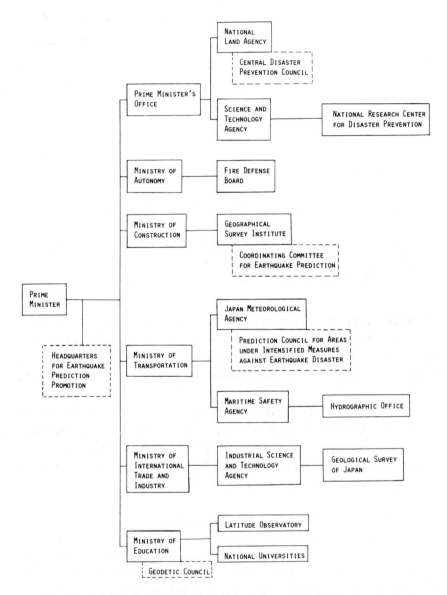

FIG. 6.2.   Diagram showing ministries and agencies and associated organs involved in earthquake prediction in Japan (SCIENCE AND TECHNOLOGY AGENCY, 1977).

TABLE 6.1. Function of various organs relevant to earthquake prediction in Japan (Science and Technology Agency, 1977 and other sources).

Organ	Function
Headquarters for Earthquake Prediction Promotion	Basic administration of earthquake prediction. Coordination of earthquake prediction work between various ministries and agencies involved.
Geodetic Council	Relistic planning of earthquake prediction program and raising proposals for the program to the government.
Sub-committee for Earthquake Prediction	Academic discussion about earthquake prediction and international liaison.
Coordinating Committee for Earthquake Prediction	Data analysis and assessment of long-term prediction.
Prediction Council for Areas under Intensified Measures against Earthquake Disaster	Short-term and imminent prediction of a great earthquake.

involved in earthquake prediction functions in Japan. In Table 6.1, the functions of various organs related to earthquake prediction administration in Japan is summarized mostly referring to the *"Binran (handbook) of Earthquake Prediction"* published by the SCIENCE and TECHNOLOGY AGENCY (1977).

Since earthquake prediction technology is still in a developing stage, it is absolutely impossible to promote earthquake prediction work only through administrative effort. The author believes that an organ such as the Headquarters should pay regard to opinions earth scientists present from scientific viewpoints.

6.1.5   Prediction Council for the Tokai area
   1)   Establishment
   As has already been mentioned in Subsection 1.1.1 and indicated in Table 1.2, a prediction council, that handles an imminent prediction, was formed in 1977 in relation to the social unrest in the Tokai area where people had become aware of the possible occurrence of a great earthquake. Being called the Prediction Council for the Tokai Area, which is a sub-organization of CCEP, the council was started in April, 1977.

In separate sheet No. 3 of the revised proposal for the 3rd 5-year program on earthquake prediction as recommended by the Geodetic Council to the Japanese government on December 17, 1976, the following is written under Item No. 3 which reads as follows;

Item No. 3. Establishment of a system for evaluation of earthquake prediction information.

Exchange of data in relation to earthquake prediction and overall evaluation based on such data have so far been made by CCEP. Although technology for short-term and imminent earthquake prediction has now been developed to some extent, as can be seen in the preceding paragraphs, however, it is now necessary to set up immediately an organ that collects and analyses relevant data almost on the real-time basis aiming at proper evaluation of earthquake prediction information connected to imminent prediction.

Although permanent establishment of such an organ must be discussed in the framework of the overall system of earthquake prediction, it should be tentatively appropriate to set up an organ that works on evaluation and judment of data taken by continuous observations in the Tokai area.

Such an organ should consist of experts having professional knowledge and considerable experience. They are to be supported by a number of staff members. In order to achieve the best function of the organ, special arrangements for the necessary budget and manpower should be made by the government.

In response to the above proposal by the Geodetic Council, the Headquarters for Earthquake Prediction Promotion decided to set up the Prediction Council for the Tokai area on April 4, 1977. The secretariat of the Council being attached to JMA, the Council was officially started on April 18, 1977. It consists of six university professors working in Tokyo. The Council usually meets every month to inspect data taken in the Tokai area.

The bylaws for management of the Council are as follows;
Objective and establishment
i)   The Prediction Council for the Tokai area is to be set up in

CCEP in order to promote short-term earthquake prediction in the Tokai area.

Organization

ii)   The Council consists of a chairman and members.

iii)   The president of CCEP takes the chair of the Council.

iv)   When the chairman is not available, a member nominated by the chairman in advance acts for the chairman.

v)   The members of the Council shall be appointed by the president of CCEP from among members of CCEP.

vi)   The term of members shall be one year. But, the members may stay on.

vii)   No substitute of the chairman and members is allowed.

Council meeting

viii)   An emergency meeting of the Council shall be called by the chairman when the observed data telemetered to JMA exhibit anomalies that exceed the limits prescribed in advance.

ix)   There is no quorum for the meeting.

x)   Should a Council meeting be called, the Headquarters for Earthquake Prediction Promotion and CCEP be immediately informed upon the direction by the chairman.

xi)   The meeting shall be held in the Section of Earthquake of JMA as a rule.

Judgment

xii)   JMA shall present all data needed for evaluation and judgment of earthquake occurrence to the Council meeting.

xiii)   On the basis of the data thus presented, the Council shall urgently evaluate and judge as to the relation between the anomalies observed and occurrence of a large earthquake in the Tokai area.

Notification of judgment

xiv)   The chairman shall immediately notify the judgment to the Central Disaster Prevention Council, Headquarters for Earthquake Prediction Promotion, CCEP and JMA.

xv)   The method of notification shall be decided separately.

Non-emergent meeting

xvi)   In order to investigate routine data necessary for judgment and other matters relevant to the Council, the chairman may call a Council meeting when necessary.

xvii)   The chairman may decide necessary matters for managing the Council after consulting at a Council meeting.

Secretariat

xviii)   Administrative business of the Council shall be handled by the Section of Earthquake of JMA.

Additional rule

This bylaw is effective as from April 18, 1977.

2)   How to make judgment

Judging from the pattern of earthquake occurrence in historical time and the spatial extent of crustal strain accumulation at present, it is surmised that the feared Tokai earthquake, should it occur, would have a magnitude of 8 or thereabout. Very few data are available for precursory effects related to this type of great earthquakes. Remember that the last earthquake struck the Tokai area in 1854 when no science of earthquake existed. It may be that the occurrence pattern of coming shock might considerably deviate from the mean tendency of earthquake occurrence for which, as discussed in Section 5.4, we may expect earthquake precursors of three classes.

It has to be assumed, therefore, that a long-term symptom might not be available in the worst case. In order to detect short-term or imminent precursors even in such a case, a 24-hour watching network has been developed over the Tokai area. The Prediction Council for the Tokai area is in a position to evaluate the data taken by the network and to make judgment as to occurrence of a great earthquake.

In order to achieve the evaluation and judgment required, all data telemeterd to JMA from observation instruments in the Tokai area on the on-line real-time basis are watched day and night. As stated in the bylaws, the chairman of the Council would be asked to come and see the data when anomalies exceed the limit prescribed beforehand. If the chairman considers that the anomalies might be connected to earthquake occurrence, all the Council members would be immediately summoned to JMA. The result of their judgment would be publicized soon after the Council meeting as shown in the flow-chart in Fig. 6.3 (SCIENCE and TECHNOLOGY AGENCY, 1977). The flow-chart having been drawn at the early stage of the Prediction Council for the Tokai area, it could be modified to some extent in

186

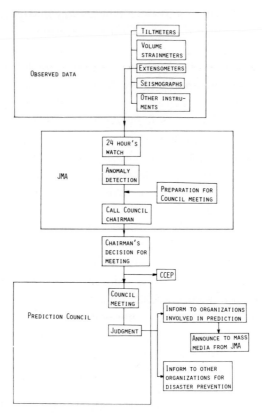

FIG. 6.3. Judgment and related flow of information of the Prediction Council for the Tokai area (SCIENCE AND TECHNOLOGY AGENCY, 1977).

view of the enactment of the Large-scale Earthquake Countermasures Act in 1978. As a matter of fact no decision on when information that 'the Council has been summoned' should be released to the public, has yet been taken.

The level of anomalies required to call the chairman has been tentatively decided as follows;

i) A volume strainmeter records a change exceeding 5×10^{-7} per 3 hours and one or more strainmeters at other places exhibit a similar change.

ii) A seismograph records 10 or more earthquakes per hour towards the west of the Suruga trough and such a state continues 2 hours or longer.

This level of anomalies was changed to a new one at the October, 1981, meeting of the Prediction Council for Areas under Intensified Measures against Earthquake Disaster as will be seen in Subsection 6.1.6.

Such a level has not been determined on a firm physical basis. As it is only for convenience of practical work, the level should be revised in the near future probably taking many more elements into consideration.

All the Council members are supposed to carry a so-called "pocket bell", a wireless alarm that can be reached by dialing a special telephone number, so that they can certainly be contacted when they are in the Tokyo area. Being university professors, however, each of them has a full-time job for research and education. It is therefore not possible for them to stay in the Tokyo area all the time. They even travel abroad from time to time when required. Judging from the importance of the Council's duty, that is concerned with the life and death of millions of people, a more powerful system for evaluation and judgment should be created.

There is no scientifically established means for actual prediction of a particular earthquake. Each Council member may have different ideas. One may rely on a probabilistic approach, while others may put much stress on an overall judgment. It is not possible to say what would happen to the judging and evaluating work when an emergency arose. It may be that the technology of earthquake prediction will some day develop to such an extent that judgment of earthquake occurrence can be effected routinely according to a mannual. In that case no prediction council is needed. The prediction work can be conducted by experts just like weather forecasting at present. The author thinks that it is the right time to start the training of such experts looking towards the further development of earthquake prediction in the foreseeable future.

The Council members are not by themselves conducting or supervising the observations that provide the data necessary for prediction. Even if they need to have a certain observation at a

---

Fɪɢ. 6.4.   The observation network for earthquake prediction in 1980, based mostly on real-time telemetering over the Tokai area.

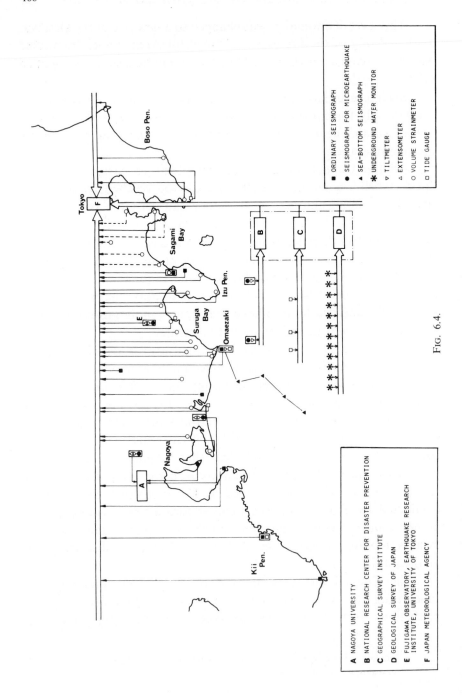

FIG. 6.4.

particular station, they have no power to order this. They may ask respective organizations to undetake that observation, but whether or not the required observation can be completed is entirely up to the organizations in question.

Even if the organizations concerned are in favor of performing such an observation, it will take time because of budgetary arrangements. At any rate it sounds ridiculous that the observation facilities are beyond the control of the Prediction Council. Such a point should be improved in the future.

3)   Observations for judgment

The observation network developed over the Tokai and South Kanto areas is shown in Fig. 6.4. The participating organizations are JMA, GSI, National Research Center for Disaster Prevention, Geological Survey of Japan, ERI, and Faculty of Science of the University of Tokyo and Nagoya University. Much of the data is telemetered to JMA via respective organizations.

The sea-bottom seismographs set up to the south-southwest of Omaezaki Point extending over 110 km must be specially mentioned. It turns out that very small earthquakes occurring off the Pacific coast of the Tokai area have now begun to be located although the seismicity is very low at present. The volume strainmeter observation was intensified in the 1980 fiscal year, i.e. the number was increased from 12 to 31.

The observation system over the Tokai area is certainly one of the most dense network for earthquake prediction in the world. As it involves observations by tiltmeter, tide-gauge, underground level-meter, and additionally by radon-meter, it is anticipated that monitoring of earthquake precursors there will undoubtedly contribute to an advance in earthquake prediction study.

6.1.6   Prediction Council for Areas under Intensified Measures against Earthquake Disaster

In association with the designation of the Tokai area as an "Area under Intensified Measures against Earthquake Disaster" on August 7, 1979 under the Large-scale Earthquake Countermeasures Act enacted on December 14, 1978, the Prediction Council for the Tokai area has been transformed to the Prediction Council for Areas under

Intensified Measures against Earthquake Disaster which is an advisory organ to JMA's Director General.

Although the Council has become an organ of a little more official character, nothing essential has been changed. All the members are still working on a part-time basis. The bylaws of the new Council are as follows;

Duty

i)   The duty of Prediction Council is as follows;

   a)   In order to carry out the duty of JMA's Director General as stated in item No. 11-2 of the meteorological service law, the Council shall evaluate and judge the occurrence of a large-scale earthquake in the Areas under Intensified Measures against Earthquake Disaster.

   b)   The Council shall conduct investigations necessary for the last clause.

Organization

ii)   The construction of the Council is as follows;

   a)   The Council consists of a chairman and members.

   b)   The chairman takes the chair of the council.

   c)   When the chairman is not available, a member nominated by the chairman in advance acts for the chairman.

Commission of the chairman and members

iii)   The chairman and members shall be commissioned by the JMA's Director General from among men of learning and experience.

iv)   The term of the chairman and members shall be one year. But, they may be reelected.

Summons and proceedings

v)   When the JMA's Director General requests the chairman to call a council meeting on the basis of item No. 8, the chairman shall at once call a Council meeting.

vi)   In addition to the case of the last item, the chairman may call a Council meeting, when necessary.

vii)   The council meeting shall be a closed one.

Request for summons

viii)   When the observed data in an Area under Intensified Measures against Earthquake Disaster exhibit anomalies exceeding the prescribed level, the JMA's Director General shall at once ask the chairman to call a Council meeting.

Judgment and report of the result

ix)   The Council summoned according to the request as stated in the last item shall immediately judge whether or not a large-scale earthquake occurs in the Area under Intensified Measures against Earthquake Disaster.

x)   The Council shall conduct judging work even after the judgment in the last item, if necessary, depending upon the development of crustal condition.

xi)   The Council shall at once report the result of judgment as described in the last two items to JMA's Director General.

Observed data and related problems

xii)   The JMA's Director General shall make arrangements for presenting data observed in the Area under Intensified Measures against Earthquake Disaster to the Council and also for making necessary government officials attend the Council meetings in order to explain the data concerned.

General affairs

xiii)   The general affairs of the Council shall be handled by the Department of Observation, JMA.

Particular items

xiv)   Necessary matters other than the bylaws here provided shall be decided by the Director of the Department of Observation, JMA on consultation with chairman.

In light of drastic increase in the number of observation stations for various disciplines in the period 1979–81, revision of the level of anomalies required to summon the Prediction Council was discussed at the monthly meeting of the Council in October, 1981. The level, which had been tentatively decided by the Prediction Council for the Tokai area in April, 1977 has been adopted as it is even after the start of the new Council.

The level of anomalies required to summon the Council was newly decided as follows;

i)   A seismograph records 10 or more swarm earthquakes per hour, including 3 or more earthquakes having magnitude 4 or over, towards the west of the Suruga trough and such a state continues 2 hours or longer along with remarkable changes recorded by 2 or more volume strainmeters at about the same time.

ii) A volume strainmeter records a change exceeding 5×10^{-7} per 3 hours and conspicuous changes of similar kind are observed by at least 3 more volume strainmeters at about the same time.

The strainmeter stations involved in the above judgment are the 16 stations located to the west of Izu Peninsula.

It was also decided that an emergency Council meeting may be summoned by the request of the chairman when he thinks it necessary.

JMA and the representative of mass media recently agreed that the information that "the Council has been summoned" is to be released to the public 30 minutes after the action for calling the Council members was taken.

6.1.7 Desirable system of earthquake prediction in Japan

As can be seen in Fig. 6.2, earthquake prediction in Japan has been promoted under cooperation between many organizations belonging to ministries and agencies different one from another. Although the cooperation centering on CCEP has so far been excellent, it appears to the author that a more unified and administratively powerful system is required for future development of earthquake prediction of practical use.

The unification of an earthquake prediction system has often been a matter of debate at the Diet. The author is of the opinion, however, that simple unification of the seismic observations of JMA, geodetic survey of GSI, and the like alone has limitations. Since all these works have long-lived traditions respectively, they function well when, and only when, they are combined with other tasks of respective organizations. For example, the seismic observation of JMA largely depends on its communication system. Someone may say, however, that a ministry of disaster prevention, which includes all the organizations such as JMA, GSI, and the like could be created. In that case, the proposed unification of an earthquake prediction system would be actualized. But the author rather doubts that such an ambitious reorganization can really be put into operation because of bureaucracy.

One of the defects of the earthquake prediction system to-date is the point that no member of CCEP or the Prediction Council is working on earthquake prediction on a full-time basis. It is the author's belief that one or two members of learning and experience should fully work on earthquake prediction all the time.

Because of these deficiencies, the author has proposed the setting up of an organization provisionally named the Headquarters of Earthquake Prediction. The organizational structure of the Headquarters will be something like the one that is shown in Fig. 6.5. The basic idea about the Headquarters is that the existing cooperative work should further be promoted. All data taken by various organizations involved would be sent on the real-time basis, if necessary, to Office of Data of the Headquarters. The data thus obtained are to be processed and analysed at the Office of Data Analysis.

In order to conduct mobile observations, a Department of Observation Work is attached to the Headquarters. The Department may set up temporary observatories, if required. It is intended for the Headquarters to handle prediction and warning at the same time, a Department of Earthquake Information which includes an Office of Public Relations should be set up. It is therefore necessary for the Headquarters to have staff members who are good at administration of this kind in addition to natural and social scientists and technicians.

The evaluation and judgment of earthquake occurrence will be the responsibility of the Director of the Headquarters. But the Headquarters will be backed up by the Headquarters for Earthquake Prediction Promotion which takes care of the budgetary and administrative side of nation-wide earthquake predition. Scientific evaluation and advice for the work of the Hadquarters will be made by an Advisory Committee for Earthquake Prediction which is similar to the existing CCEP.

The author thinks that a Department of Researches should be included in the Headquarters in order to keep up with ever-developing technology of earthquake prediction and to secure eminent researchers. It is the author's guess that only 100 members or so would be necessary for starting the Headquarters.

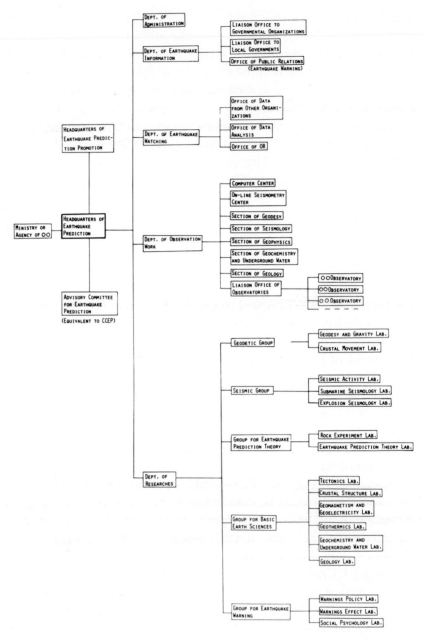

FIG. 6.5. Organizational composition of the proposed Headquarters of Earthquake Prediction (tentatively named).

The above idea being only the author's personal one, much debate should be allowed for actually setting up an organ which superintends nation-wide earthquake prediction work. In any case we hope that a headquarters, center and the like, whatever it may be called, should be established as soon as possible, otherwise no dramatic development of practicable earthquake prediction would be possible in Japan.

6.2   System of Earthquake Prediction in the U.S.A.

A sort of earthquake prediction council consisting of 5 members has been set up in the USGS. Although it has been proposed to form a National Earthquake Prediction Council by adding members from the outside of USGS, such a council could not be created until recently because of a regulation that summons of a committee of federal government should be announced at least 15 days before the meeting (WORKING GROUP ON EARTHQUAKE HAZARDS REDUCTION, 1978). According to R. L. Wesson (personal communication, 1979), such a ridiculous regulation for the council that handles an imminent matter such as earthquake prediction has somehow been avoided and so a national council with 15 members was set up in 1979. The council is chaired by someone not working in USGS.

It is the author's understanding that the council is a kind of evaluation council, so that no prediction is made by the council itself. Being named the U.S. National Earthquake Prediction Evaluation Council, it rejected the prediction made by B. D. Brady, a physicist at the U.S. Bureau of Mines on January 27, 1981. Brady has been insisting that an extremely large earthquake would hit Preu some time in August 1981. The prediction has caused a considerable unrest among Peruvian people (KERR, 1981). The eathquake in fact did not occur, and Brady had withdrawn his prediction some time in May.

The Governor of California has his own California Earthquake Prediction Evaluation Council. When a prediction is put forward by USGS or someone-else, the Governor asks the Council to evaluate the prediction in question. Should the result of evaluation be positive, the

Governor would take necessary measures for mitigating the earthquake disaster.

In 1976, the Council was called in relation to the southern California uplift (Fig. 1.19) which had been officially notified to the Governor by the USGS director. As a result of the couclusion at the Council meeting, the information about the uplift was transmitted to the locale along with the necessary countermeasures in case of emergency. The Council met again in the same year and evaluated a report in newspapers concerning possible occurrence of a large earthquale in the Los Angeles area, the report having been based on the hypothesis of a scientist. The Council discredited such a report.

No nation-wide system of earthquake prediction such as those in China and Japan is in operation in the U.S.A. probably because of the seismicity lower than that of the above two countries. However, an extensive array of various disciplines aimed at earthquake prediction has been developed mainly over California by USGS, California Institute of Technology and other organizations.

6.3 System of Earthquake Prediction in China

6.3.1 State Seismological Bureau
The State Seismological Bureau (SSB) is responsible for nationwide observation and prediction of earthquake in China. SSB being attached to the State Council, which is equivalent to the Cabinet in Japan, was set up in 1971 in order to unify administration, observation and research in relation to earthquake. Until that time those duties had been handled by a number of ministries of State Council.

SSB consists of about 2,000 members, 60% of which are scientists and technicians and the remaining 40% are administrators. Large sections of Geophysical and Geological Institutes have been transferred to SSB in 1978 as already stated in Section 1.4, so that it is said that about 400 members have moved to the new institutes in SSB.

The main duties of SSB are as follows;
i)   Designation of areas where a large earthquake seems likely to occur at national meetings for investigating all China seismic activity
ii) Establishment of fundamental policy for earthquake disaster

prevention and rescue work
- iii) Promotion of basic and applied researches on earthquake
- iv) Collection and preservation of data for earthquakes
- v) Direction of Province, County and other local seismological bureaus and synthesis of earthquake prediction

6.3.2 Province seismological bureau, brigade and team

Province seismological bureaus are the centers of earthquake prediction in respective provinces. They are responsible for standard seismic observation, maintenance of seismic observation networks, geophysical observation of various disciplines, determination of seismicity in respective areas, manufacturing and distribution of earth-current meters, tiltmeters and other instruments, announcement of earthquake prediction and the like.

When the activity is not as extensive as to be called a bureau, such a center of earthquake work is called a brigade. Although a seismological brigade is at the moment in operation in Peking (Beijing), a city under the direct control of central government, it will shortly be promoted to a bureau. A unit, the scale of which is smaller than that of a brigade is called a team.

It is said that 300, 700, and 600 members are working respectively at seismological bureaus of Liaoning, Yunnan and Sichuan Provinces. According to the SHIZUOKA PREFECTURE (1978), the organization and manpower of the Yunnan Seismological Bureau are shown in the following in which the numerals indicate the number of people working on the subject:

- i) Seismic observatory and observation point—200
- ii) Analysis and prediction office—50

   To work on information analysis, medium- and short-term and imminent predictions. Analysis is usually made once a week. The result is reported to the Province Revolutionary Committee and SSB. The office is also responsible for public relations in the province and exchange of data and evaluation between neighboring provinces.

- iii) Geological party—50

   To work on active faults and long- and medium-term predictions.

iv) Geodetic survey party—80
   To work on levelling and triangulation surveys.
v) Factory of instruments for observation—80
   a) Design and production of instruments for amateur use
   b) Repair of professional instruments
   c) Mobile party for repairing
vi) All-round experimental research institute—40
vii) Party for transportation—70
viii) Computer center—10
ix) Other work including libraries, collection of data and information, supply, general affairs, welfare, and so on—120

6.3.3 Earthquake office

Earthquake offices, which are attached to revolutionary committees or people's governments of various levels, take care of administration in regard to earthquake. Earthquake information is communicated to revolutionary committees through the offices based on the analysis made by seismological bureaus and brigades. The offices are also responsible for promotion of spread of earthquake knowledge.

6.3.4 Standard seismological observatory

As has already been stated in Section 1.4, there are 17 standard seismological observatories in China. Seismic, geomagnetic and geoelectric, crustal movement, gravity and geochemical observations are carried out there. The Peking (Beijing) Standard Seismological Observatory at Paichiatang (see Fig. 1.32) belongs to the Geophysical Institute of SSB, while other standard observatories are attached to province seismological bureaus.

6.3.5 Seismological observatory, seismic observation point and mobile station

A seismological observatory works as a center of earthquake observation at a locale. Sometimes consisting of more than 10 members, a seismological observatory also works on data analysis and education of the public at large. It is said that there are about 300 seismological observatories all over China.

In addition to such observatories, there are many observation points equipped with limited apparata. In case of emergent activity, numerous mobile stations are usually set up.

6.3.6   Block-diagram of the Chinese system of earthquake prediction

The Chinese system of earthquake prediction is organized under the close cooperation between the various organizations as mentioned above and central and local administrative units. Figure 6.6 indicates a rough idea of the Chinese system although the detail of the system is so complicated that it is hard to show it very clearly in such a diagram.

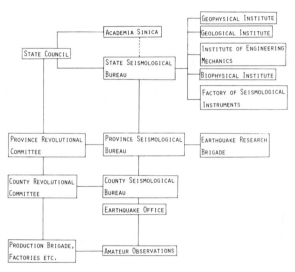

FIG. 6.6.   Block-diagram of the Chinese system for earthquake prediction (Shizuoka Prefecture, 1978).

When the first U.S. seismological mission visited China, they were surprised to see that several hunderd scientists and several thousand technicians and workers were working at about 250 seismological observatories and about 5,000 observation points (ALLEN et al., 1975). According to SHIZUOKA PREFECTURE (1978), however, the number of people working on earthquake prediction (including administrators) exceeds 10,000 and that of amateurs amounts to about 150,000. It should therefore be said that earthquake prediction work is fantastically extensive in China.

CHAPTER 7

COMPARATIVE STUDY OF EARTHQUAKE PREDICTION

It is interesting to compare the approaches to earthquake prediction between the major four countries, i.e. Japan, the U.S.A., China, and the U.S.S.R. Since there are differences in the pattern of earthquake occurrence, the conditions of country, the national traits and the social structure between these countries, it may be natural that the approaches are different from nation to nation.

The occurrence pattern of extremely large earthquakes in Japan, Kamchatka, Aleutian, and Alaska is characterized mostly by reverse faults at subduction zones. Earthquakes related to the San Andreas fault in California are largely of strike-slip type although vertical slips become apparent in the neighboring areas such as Nevada and Utah. Large earthquakes in the continent of China are almost of shallow origin accompanied with much earthquake disaster. In Japan, inland earthquakes having a magnitude of 7 or thereabout are also highly hazardous.

In contrast to the long history in China and Japan, the seismic history in the U.S.A. covers only 200 years or so, an extremely short period compared to those in the above two countries. It cannot be said, therefore, that even the eastern U.S.A., where the seismicity has been very low in recent years, is absolutely safe from earthquakes. In China, revivals of seismic activity have been noticed in a number of cases several hundred years after an active period that had also spanned several centuries. As pointed out by the first U.S. seismological mission to China (ALLEN et al., 1975), a possibility of uprise of seismic activity in the eastern U.S.A. cannot be ruled out.

Earthquake prediction programs have been started in the four major countries essentially in the 1960's. It is really surprising that the Chinese program, that started from a practically zero level, has developed to the present extent in such a short period of time. As for the annual budget for earthquake prediction programs, the Japanese program almost doubles the U.S. one because salaries and overheads are included in the latter. It appears to the author that the American budget could be substantially increased judging from the national power of the U.S.A. It should be pointed out, however, that it took more than 10 years to reach the present budgetary level of the program in Japan.

It is hard to estimate the budget of the Chinese program in western currencies because of the difference in the social conditions and prices of commodities between the western countries and China. Although it was once mentioned that the Chinese budget of earthquake prediction work is in the order of million dollars (PANEL ON EARTHQUAKE PREDICTION OF THE COMMITTEE ON SEISMOLOGY, 1976), it appears to the author, who visited China in 1978, 1980, and 1981, that the Chinese budget of earthquake prediction work reaches almost a level comparable to those in Japan and the U.S.A. provided the difference in the prices of commodities between China and the western countries is taken into account.

In Table 7.1, the author attempts to make a comparison of various items involved in earthquake prediction between the four major countries. Looking at Table 7.1, it is interesting to note that items on which respective countries are putting emphasis can be clearly indicated in the table. The Japanese program traditionally attaches importance to grodetic survey, so that probabilities of an earthquake occurring in a certain area can sometimes be estimated. Geodetic work comparable to the Japanese system is underway over the San Andreas fault in California, U.S.A. Detection of ground tilting by levellings over a short distance is fashionable in China. Effort toward detecting an anomalous crustal deformation relative to the sea level by means of tide-gauge observation is intensively made in Japan.

TABLE 7.1. Comparison of earthquake prediction work between Japan, the U.S.A., China, and the U.S.S.R.

Item		Japan	U.S.A.	China	U.S.S.R.
Type of large earthquakes		Inter- and intra-plate	Almost inter-plate but some intra-plate	Intra-plate except Taiwan	Inter-plate (Kamchatka) and intra-plate (Middle Asia)
Data coverage		ca. 2,000 yr	ca. 200 yr.	ca. 3,000 yr	?
Earthquake prediction program	Inauguration	1965	1973 (preliminarily during 1960's)	1966	Later half of 1950's
	Manpower	500	500	Professional: 10,000 Amateur: 150,000	1,000
	Yearly budget	ca. 6×10^9 yen	ca. 1.5×10^7 dollars	Equivalent to those of Japan and the U.S.A.	ca. 2×10^7 roubles
Earthquake prediction elements	Geodesy	Put much stress on trilateration and levelling	Put much stress on electro-optical survey over San Andreas fault	Put much stress on short levelling	Levellings are conducted in Garm area
	Tide-gauge	About 100 stations every 100 km along coast	Less dense than Japan	Less dense than Japan	—

TABLE 7.1 (continued)

Item		Japan	U.S.A.	China	U.S.S.R.
Earthquake prediction elements	Gravity	Detection of gravity change associated with land uplift	The same as Japan	Detection of changes larger than 100 μgal	—
	Continuous observation of crustal movement	Popularization of volume strain-meter	Bubble tiltmeters over San Andreas fault	Agreement between results of short levelling and water-tube tiltmeter	Precursory changes in Middle Asia
	Seismic activity	Real-time sea-bottom and deep-well seismographs	About 500 seismic stations with telemetering in California	Telemetered network around Peking (Beijing)	—
	Seismic wave velocity	No remarkable change	No remarkable change	Many reported changes without discussion about accuracy	No new finding
	Earth tide	Detection of precursory change	—	—	Detection of precursory change
	Geomagnetism and earth-currents	Significant changes in Izu Peninsula	—	Magnetic changes preceeding Tangshan and Songpan-Pingwu	—

TABLE 7.1 (continued)

Item		Japan	U.S.A.	China	U.S.S.R.
Earthquake prediction elements	Geomagnetism and earth-currents			earthquakes. Tele-metered proton precession magne-tometers around Peking (Beijing)	Use of MHD generation
	Resistivity	Detection of pre-cursory effect by Yamazaki resistivity variometer	—	Many examples of precursor	—
	CA transfer function	promising as precursors	—	A few precursory effects	
	Underground water level	A few precursors	—	Many precursors	Precursory changes in Middle Asia and Kurile Islands
	Randon, helium, and other geochemical elements	Promising as precursors	Observation over San Andreas fault	Important element of prediction	Very popular in Middle Asia
	Fire ball	—	—	Many instances in Sichuan and Liaoning Provinces	—

TABLE 7.1 (continued)

Item		Japan	U.S.A.	China	U.S.S.R.
Earthquake prediction elements	Anomalous animal behavior	Some possibility	Some possibility	Important element of prediction	No recent reports
System for prediction		CCEP and Prediction Council for Areas under Intensified Measures against Earthquake Disaster	National and USGS Prediction Councils	SSB and Province Seismological Bureaus	Academies of respective Republics
Strategy of prediction		Overall judgment with probabilistic consideration	Not clear	Overall judgment putting stress on macroscopic anomalies	Multi-discipline observation
Actual results of prediction		Forecast of Nemuro-Hanto Oki earthquake at Diet about 2 months earlier	Non-official predictions of magnitude 4–5 earthquakes	Success in predicting 4 earthquakes of magnitude 7 or over in 1975–1976	Success in predicting earthquakes of magnitude 7 in Pamir and Iran in 1978
Amateur participation		Not much	Not much	Very extensive and useful	—
Research on public reaction against a prediction		Unsatisfactory	Completed preliminary studies	Nothing	Nothing
Legislation		Large-scale Earthquake Countermeasures Act in 1978	Earthquake Hazards Reduction Act of 1977	—	—

Gravity changes reported from China are incredibly large, while there is a growing tendency to discuss those, mostly associated with land uplift and subsidence, in relation to earthquake prediction in Japan and the U.S.A.

Extensive telemetry of microrarthquake observation has been developed in Japan, California and the Peking (Beijing) areas. Accordingly, computer processing of seismic data has now become fashionable. The real-time system of sea-bottom seismographs established some 100 km off the pacific coast of the Tokai area and the system of deep-well seismic observation for which the depth of seismographs reaches 2,000–3,000 m would have been impossible if the earthquake prediction program in Japan had not been developed extensively.

Volume strainmeters originally developed in the U.S.A. are now widely used by JMA in the hope of providing a powerful means for imminent prediction of the feared Tokai eartquake. Bubble tiltmeters have also been widely used along the San Andreas fault although some scientists throw doubt upon the accuracy of the observed results because the instruments are set up in relatively shallow holes.

Many geomagnetic and geoelectric precursors observed in China prove defective because of a lack of measuring units on recording paper, no estimate of errors and so on. The author hopes that these points should soon be improved in such a manner that international comparison of data becomes possible.

The Yamazaki resistivity variometer in operation at a station 60 km south of Tokyo has been observing precursory effects for many large and moderately large arthquakes in the last 10 years. Only very few outstanding precursory changes in resistivity have been found by the observation along the San Andreas fault partly because of low seismicity there. Soviet resistivity work seems to have had success in monitoring precursory effects.

Pursuing precursory anomalies of underground water are popular in the Chinese and Soviet programs. Many examples of premonitory changes in underground water level and in chemical constituents dissolved in underground water have been put forward. Researches in this line have also commenced in Japan and the U.S.A. in recent years.

Much attention has now become drawn to the so-called macroscopic phenomena as described in Section 4.10. Appearance of fireballs forerunning an earthquake as reported from the Sichuan Province, China, sounds fantastic. Anomalous animal behavior has now become one of the disciplines of earthquake prediction programs in China, Japan, and the U.S.A.

The usual approach to earthquake prediction in the U.S.A. seems likely to suppose a model of source process and to try to observe phenomena likely to be associated with that model. In contrast to such an approach, colleagues in China (and probably in the U.S.S.R.) try to observe everything which might have something to do with some precursory effect. The Japanese approach may be positioned somewhere between these two extremities. It is hard to judge which approach is better at present. Although the U.S. approach is very efficient provided the model should be a good approximation, there may be some instances for which a more broad-minded approach is required. The Chinese way of approach may sometimes be so inefficient that it takes much time to achieve some success.

In socialistic countries such as the U.S.S.R. and China, no serious problem arises in the case of actual prediction of a large earthquake because the political and administrative control of government is usually very strong. On the contrary, many difficulties would certainly be met by central and local governments in capitalistic countries such as Japan and the U.S.A., should an earthquake warining actually be issued. As people would ask the governments to compensate economic losses due to the issuance of prediction and warning, far-reaching measures should be implemented for controlling possible economic difficulties. Psychological panic would also occur. It is therefore necessary to formulate special legislation in order to deal with these points. Actually the Large-scale Earthquake Countermeasures Act and Earthquake Hazards Reduction Act of 1977 have been enacted respectively in Japan and the U.S.A.

Study of public reaction against an earthquake warning is best developed in the U.S.A. It is important to promote this kind of study in order to be aware of the best timing of warning issuance. It is regrettable to hear that the 1979 research funds of NSF for this kind of study has been cut down for some reason (DICKSON, 1979).

One of the most outstanding characteristics in the Chinese program on earthquake prediction is the fact that countless amateurs are participating to the observations as volunteers. Even in the U.S.A. and Japan, efforts toward organizing amateur observations by interested citizens has recently started.

CHAPTER 8

EARTHQUAKE WARNINGS —CASE HISTORIES

8.1 Matsushiro Earthquakes

A set of NOAA's standard seismographs was installed at the Matsushiro Seismological Observatory of JMA in the beginning of August, 1965. Being located nearby Nagano City in Central Japan, the observatory is famed for its large-scale and well-equipped underground vaults which were capable of use as a shelter for the Japanese royal family towards the end of World War II. It was originally intended by JMA to make a contribution to the worldwide network of seismic observation by introducing the standard seismographs there, so that emphasis was put on the observation of long-distance earthquakes.

Ironically enough, however, the seismographs at the observatory began to record microearthquakes occurring nearby the observatory from August, 1965.

The number of microearthquakes had increased so steadily that the standard seismographs at the observatory tended to record several hundred shocks a day. The number of felt earthquakes also increased substantially. On November 4, an earthquake, of which the ground motion intensity estimated as IV according to the JMA's intensity scale (equivalent to intensity V in the modified Mercalli scale), hit the Matsushiro area for the first time. Following that shock, the seismic activity kept its level with a slightly increasing tendency, so that three intensity IV earthquakes occurred successively during the evening of November 22. Such a state of seismic activity continued until March, 1966 with occasional occurrences of shocks that recorded intensity V (equivalent to intensity VII in the modified Mercalli scale).

From the middle of March, 1966, the number of earthquakes began to increase enormously day by day. The number of earthquakes recorded at the Seismological Observatory exceeded 6,000 a day, while more than 600 earthquakes were felt by the local inhabitants. JMA, ERI, and other organizations had of course been monitoring the seismic activity in the Matsushiro area. On the basis of the observed results of various disciplines including seismic activity and crustal movement, the overall view of JMA and ERI on the seismic state in the northern Nagano Prefecture was released on April 2, 1966, and the possibility of further violent seismic activity in the area and its expansion into the neighboring areas was announced to the public. Actually Nagano City was hit by an earthquake of intensity V on April 5, for the first time during that sustained activity. The activity seemed to reach its maximum on April 17 when three shocks of intensity V and three shocks of intensity IV were successively recorded.

In association with the swarm activity of earthquakes in the Matsushiro area, it was quite natural that the public demanded frequent information from the authorities on the up-to-date state of seismic activity and its growth or decay in the immediate future. In order to reply to such social demands, the Committee on Crustal Activity related to the Matsushiro Earthquakes was established under a resolution of the Section of Earthquake Prediction of the Geodetic Council towards the end of April (see Table 1.2). The committee consisting of university professors and experts working in JMA, GSI, and so on, much of the growth and decay of seismic activity, crustal movement and the like was analysed by this committee. Based on the conclusions of the committee, earthquake information was released to the public through the local office of JMA. It can be said that the committee was the prototype of CCEP established in the later year.

Earthquake information released to the public based on discussions and conclusions reached by the committee, followed this general release "The probability of a slightly damaging earthquake occurring somewhere around so and so city, town or village within a few months is high". This is essentially an earthquake warning. Koshoku City, Togura, Sanada and Kamiyamada Towns, and Sakai and Azuma Villages were the targets for such warnings, and all these areas were hit by earthquakes of magnitude 5 or so sooner or later as forecasted. Although the announcements disclosed by JMA were called earthquake information, they were actually earthquake warnings publicized for the first time in history on the scientifically credible basis.

Reactions of local people against the earthquake information varied depending upon profession, living condition and so on. Local governments behaved in general in a highly responsible manner. Upon receiving the information, they worked hard on mending school buildings, augumenting fire brigades and so on. On the other hand, those operating hotels, inns, and sightseeing businesses were said to have complained because of the inevitable drop in the number of guests.

The reasons why the local inhabitants took a calm attitude in spite of earthquake warnings may be twofold. First of all they were accustomed to earthquakes, having previously experienced thousands of shocks during the most violent period of seismic activity. The point that the expected magnitude of predicted shocks would be of the order of 5 at most was also relevent. Judging from the extent of area in which earthquake swarms were taking place, it was highly probable that no earthquake of which the magnitude exceeds 6 would occur. This point had been emphasized and publicized by seismologists from time to time, so that the local people tended to believe that no devastating shock would hit after all.

The activity of Matsushiro earthquakes, which became violent in April, 1966, revived again in August in the same year although it tended to become gradually quiet in the following years. Towards the last peak of activity, geophysical activities such as crustal upheaval exceeding 70 cm, upwelling of much water, increase in th geomagnetic field as large as 10γ and the like were observed although nothing about those geophysical events will be dealt with here.

As a result of the Matsushiro experience, it is the author's belief that an earthquake warning could sometimes be very useful for preventing earthquake damage and keeping the locale in order. It has been brought out, however, that the timing and style of earthquake information issued must be well-organized. To the surprise of the seismologists, it was sometimes rumored that seismologists knew that a strong earthquake would shortly occur, although they did not inform the public. It was also experienced that explanations by seismologists was not communicated to the public factually and that the flow of information was modified incorrectly. This was because of misinterpretation by pressmen not experienced in earthquake phenomena. At any rate, the Matsushiro experience makes it clear that officials who are responsible for earthquake warning and local people who are going to receive the warning should be trained for earthquake prediction more extensively.

In the case of the Matsushiro earthquakes, there was no suspension of electricity, communication and traffic, so that public relations activity proceeded without interruption. The situation would be quite different for an earthquake of much larger magnitude without long-continued swarm activity. In such a case, it is anticipated that an actual warning issuance would certainly meet various difficulties.

8.2 Haicheng Earthquake

As was mentioned in Subsection 2.1.2 and shown in Fig. 2.1, seismic activity along a system of faults running in a southwest-northeast direction has been resumed in the Hebei and Liaoning Provinces in recent years. Following the Xingtai earthquakes ($M = 6.8$, 7.2, 1966), there occurred the Hejian earthquake ($M = 6.3$, 1967) in Hebei Province and the Bohai Wan earthquake ($M = 7.5$, 1969). On taking such a tendency of epicentral migration into consideration very seriously, the southern part of Liaoning Province was designated an area of accentuated observation at the 1st National Conference on Earthquake Prediction in 1970. This was actually a long-term prediction.

The Earthquake Office of Lianoning Province was then established in March, 1970 in order to promote unified countermeasures against coming earthquake. (The office became the Province Seismological Bureau later in 1975 after the Haicheng earthquake.)

At the same time, geoscientific obsevations of various disciplines were so intensified that 14 seismological observatories and observation points were provided. Anomalies of crustal deformation, geomagnetism, and sea level were found by 1974 as stated in Chapter 4. As can be seen in Fig. 4.17, the frequency of earthquake occurrence in the southern part of Liaoning Province increased by about a factor of 5 compared to a normal period.

Based on the above data, the possibility of violent earthquake occurring in the northern Bohai area within 1–2 year's time was advised at a regional conference for the North China-Bohai area convened by SSB in June, 1979. Accordingly, the Revolutionary Committee of Liaoning Province issued a medium-term earthquake prediction.

At this stage, geophysical and geochemical observations were intensified. At the same time education of the public about earthquakes was emphasized in the southern Liaoning Province. The public relations activity aimed at increasing the knowledge about earthquakes among local people was intensified. The stressed items are as follows; i) Why does an earthquake occur? ii) Earthquake prediction, especially the use of macroscopic phenomena such as animal anomaly, change in well-water and the like. iii) How to make houses earthquake-proof? For the purpose of promoting such public relations activities, a number of means such as motion pictures, slides, pamphlets, exhibitions, lecture meetings, and so on are utilized. Such acitvities were developed through towns and people's communes. It was hoped that the understanding of earthquake phenomenon would prove useful in case of short-term and imminent earthquake prediction.

When the author's mission visited a production brigade of a commune in the epicentral area of the Haicheng earthquake in 1978, Sun Ya-jie, the vice-leader of the brigade, told us that even the eldest villagers had never experienced an earthquake, so that knowledge about earthquakes was completely lacking among villagers. Such defects were only covered by the education and propaganda by the Communist Party. The situation is completely different from that of Japan where we feel moderately large and small earthquakes from time to time.

In November, 1974, a conference for the three northeast provinces was held. According to the data around this period, it was guessed that the activity of the Jin Xian fault was getting high from the results of short levelling, geomagnetic observation and microseismicity. As a result of discussion at the conference, it was concluded that there was a possibility of a destructive earthquake occuring in the Yingkou-Dalian area in the near future. Accordingly, short-term prediction information was issued by the Liaoning Province Revolutionary Committee.

Long- and medium-term predictions are not in general issued to the public at large in China. They are communicated only to administrative organizations of various classes. It may be that a warining sent out too early is considered to cause needless confusion. Information to the public is disclosed only at the stage of short-term prediction, when action for preventing earthquake disaster is deemed necessary and to be taken immediately. In a country like Japan where mass means of communication has developed extensively, the author is of the opinion that we cannot follow the Chinese example. Earthquake prediction information, whether it is for long-term or short-term, cannot be kept secret over a long period of time because the press would sense out the story sooner or later.

Towards the middle of December, 1974, many outstanding reports on macroscopic observation began to reach the authorities in Liaoning Province. For example hibernating snakes came out on the ground covered by snow only to die in the freezing weather. A group of rats appeared in the Dandong area, they were so panicked that they did not pay attention to man. A man at a people's commune in that area found many rats, 20 or more in number, and he could catch them easily with his hands. Remarkable anomalies were also observed by geophysical instruments at almost the same period. For example, the levelling survey along short base-lines across the Jinzhou fault at Jin Xian brought a reversal of tilting direction to light at the end of December, 1974, as was already mentioned in Section 4.3 and illustrated in Fig. 4.4. A similar ground tilting from the northwest to the southeast was recorded at the Shenyang Seismological Observatory.

It appears that there was some confusion for the earthquake

countermeasures administration in the Liaoning Province towards the end of the year 1974 because such an administration was in process of organization and development. It sometimes happened that earthquake warnings were presented to the locale separately by each seismological observatory. An order of evacuation was issued in the Panshan area on December 28, 1974, so that 20,000 to 30,000 inhabitants took refuge outside their houses in the freezing weather although no large earthquake occurred. Just after the short-term prediction on December 20, an earthquake of magnitude 4.8 occurred somewhere between Liaoyang and Benxi about 70 km north of Haicheng. But it is hard to say that the time and place of occurrence of the quake was predicted.

SSB called the 2nd National Conference on Earthquake Prediction in the middle of January, 1975. It was concluded at the conference, which looked into all the available data in detail, that an earthquake of magnitude 6 or thereabout seemed likely to occur somewhere around the Yingkou-Jin Xian area some time in the early half of the year 1975. It was reported that anomalies of animal behavior, gush of underground water, rising or lowering and bubbling of well water and so on were prevailing all over the Liaoning Province at this period. The seismic activity over the Jin Xian-Yingkou became quiet forming a seismic gap.

The Province Earthquake Office convened a meeting of representatives of each city and seismological observatory on January 28. They discussed emergent arrangements for earthquake disaster prevention, relief activity, safety of dams, railways, electric power plants, dangerous articles, and other buildings. Propaganda about earthquake disaster prevention was more intensified resulting in a much higher public concern. Accordingly, the number of reports on macroscopic anomaly increased. Exercises for evacuation and anti-earthquake measures were preformed at coal mines, factories and highly populated areas.

On January 30, 1975, the ground tilting at the Shenyang Seismological Observatory changed its direction from the southeast to the southwest, the tilting speed having also accelerated. The interior meeting of the Observatory on January 31 concluded either that an earthquake of magnitude 5 or thereabout would occur in the

Liaoyang area or that an earthquake of magnitude 6 or thereabout would occur in the Jin Xian-Gai Xian area in the immediate future. This conclusion was reported to the Province Earthquake Office.

Microearthquakes began to occur at the prefectural border between Yingkou and Haicheng on February 1, 1975. Anomalous animal behavior became so marked that pigs bit each other at a pigpen in the Panjen area on February 2, so that 10 or more baby pigs lost their tails. Many pigs lost their appetite and climbed up on hedges and gates.

As can be seen in Fig. 4.30, the earth potential observed by the 102 Brigade for Geological Prospecting at a station near Haicheng recorded a sudden decrease on February 2. On the next day, frequency of microearthquakes increased enormously, reaching 20 shocks an hour towards evening. It was reported from production team in Yingkou that 4 cows began to fight each other and another 2 cows scratched the ground at about 6 o'clock in the evening on that day. The Province Earthquake Office analysed these data from all parts of the province arriving at the conclusion that occurrence of a large earthquake would be highly probable after the micro-seismicity had taken place in the Yingkou-Haicheng area at 00 : 30 on February 4. This conclusion was immediately conveyed to the Province Revolutionary Committee.

The 102 Brigade for Geological Prospecting began to observe strong pulses in earth potential, so that observation became difficult to carry out from time to time from around 16 : 00 on February 3. Towards 11–12 o'clock in the evening, the voltage between the north-south electrodes exceeded 1 volt. Water spurted out from a hand-pumping well, as shown in Fig. 1.39, at a village in the focal area at about 8 o'clock in the morining. Around noon on that day, water began to splash at a height of 1 m.

Microearthquake activity increased not only in frequency but also in magnitude of individual shock. Before noon on the morning of February 4, earthquakes of magnitude 4.2 and 4.7 occurred. But, the activity suddenly decreased as can be seen in Fig. 4.11.

The Province Revolutionary Committee then issued an imminent earthquake warning and instruction for earthquake disaster prevention to the whole provincial area. The Province Earthquake Office moved to Haicheng in oredr to cope with the earthquake expected in the Yingkou-Haicheng area at 14 : 00 in the afternoon.

Many deer at a breeding farm in the Anshan area became agitated at 10 : 50 in the morning on February 4. They jumped and ran madly. When they ran away breaking down the gate, a three year old deer broke its legs as the herd pushed and shoved. In the village where water gushed out from a well shown in Fig. 1.39, ducks flew over a distance of 100 m and well water became muddy in the afternoon. Upon receiving the imminent instruction from the upper class committee, each revolutionary committee in the Yingkou-Haicheng area convened emergency meetings and directed local inhibitants to move to shelters and to take cattle out of pens.

The Shihpengyu Seismological Brigade which is quite close to the epicenter gathered the local people in a square where they were shown motion picture films in the evening. When the 2nd film was being shown on the screen, the Haicheng earthquake of magnitude 7.3 occurred at 19 h 36 m 06 s. However, very few people were killed because of various countermeasures successfully taken.

Geophysical and geochemical observations aimed at earthquake prediction have been continued or rather intensified in Liaoning Province after the Haicheng earthquake. When the author visited the Haicheng area in 1978, he was told by members of the Province Seismological Bureau that the southern part of Liao-tun Peninsula is still under a long-term prediction.

An earthquake of magnitude 6.0 occurred at Guantun on May 18, 1978. The focal depth having been 8–10 km, building walls and hedges were damaged. A number of people were injured by broken pieces of glass. In relation to this earthquake, a medium-term prediction had been put forward as a result of the National Conference in February, 1978. It was surmised that an earthquake of magnitude 6 or thereabout would occur in southern Liaoning Province. Based on anomalies of the short levelling at Jin Xian, microseismicity, radon concentration and underground water level, the Earthquake Office of Yingkou City issued a short-term prediction towards the end of April that a magnitude 6 earthquake might occur in the Yungkou-Jin Xian area some time between May 20 and 31. However, no imminent prediction was issued. The shock occurred a little to the north of the suspected area, while no extremely short-range precursors were observed.

8.3 Tangshan Earthquake

Tangshan, that is located about 150 km east of Peking (Beijing), is famous for coal mining and heavy industries such as steel, chemicals, electric machinery, electronics, and so on. It was said that the population was more than one million. A magnitude 7.8 earthquake hit the Tangshan area on July 28, 1976. According to a non-confirmed report, there were 655, 237 dead and 780,000 injured. However, the Kyodo Press Service reported from Peking (Beijing) on November 22, 1979, that it was officially announced by the Chinese authority that 242,000 plople were killed and 164,000 severely injured by the quake.

The author and his missions were not allowed to enter the Tangshan area until 1981. This was also the case for other missions. The following is a summary of what the author was told by members of Hebei Province Seismological Bureau and assessed from other sources.

It is widely known that no imminent earthquake warning was issued prior to the Tangshan earthquake. It should be borne in mind, however, that long-, medium- and short-term predictions had been made on the basis of ample data. That a regional conference, convened by SSB in June, 1974, concluded that a destructive earthquake would occur in the northern Bohai area as was stated in the last section. The conference also concluded that there was a possibility of an earthquake of magnitude 5 to 6 occurring in the Peking (Beijing)-Tianjin-Bohai area within a few years' time. The conclusion was based on the anomalies in levelling, gravity and underground water. The tendency of westward migration of seismicity as mentioned in Subsection 2.1.2 was also taken into account.

As for the medium-term prediction, Chinese colleagues paid attention to the changes in ground tilting as revealed by the short levelling and the water-tube tiltmeter observation at Dahuichang and Niuk'ouyu as shown in Fig. 4.8 and discussed in Section 4.3. This sudden change started in the middle of the year 1975 seems likely to have somethig to do with the earthquake occurrence. In addition to the premonitory ground tilting, decreases in *b* value (LI *et al.*, 1978; Fig. 4.19) and earth resistivity (ZHAO and QIAN, 1978; Fig. 4.33) were detected. NORITOMI (1978a, b) reported on a number of resistivity changes, that seem to have forerun the Tangshan earthquake, as observed at other stations. WAKITA (1978a) reviewed precursory changes in radon concentration and concluded that some increase (by 20% on the average) had taken place at 20 wells within a distance of 200 km from the epicenter. According to ZHENG *et al.* (1979), the level of well-water at scores of wells in the Tangshan-Tianjin area tended to lower considerably several years before the earthquake. The level at Baitangko had lowered by 40 m since 1970. It seems likely that such a lowering of well-water was caused partly by the drought over a long period and partly by the accumulation of crustal strain related to the earthquake occurrence.

On the basis of these anomalies, SSB instructed the ministries of State Council involved that a destructive earthquake seemed likely to occur in Tianjin-Peking (Beijing)-Tangshan area within 2 years' time, so that every measures to prevent earthquake disaster should be taken at factories and other facilities. According to OIKE (1978b), earthquake damage to factories and the like thus became very much lessened in spite of the failure of imminent warning.

In January 1976, the Hebei Province Seismological Bureau held a conference where the seismic state in the Hebei-Liaoning area was argued. But the area, regarded as dangerous, was so wide that no precise area, which would become the target of imminent warning, could be specified.

It was noticed towards April, 1976, that the above-mentioned anomalies were accelerated or reversed. A magnitude 6.3 earthquake occurred about 300 km to the west of Peking (Beijing) on April 6, and another one of magnitude 4.4 occurred about 70 km to the southwest of Tianjin on April 22. In June, microearthquakes began to occur in a swarm at Dayuan, Shandong Province, to the southern Bohai Wan. However, the Tangshan-Tianjin area was still in a state of seismic gap.

In June, SSB called a National Conference on Earthquake Prediction where the above view of the Hebei Province Seismological

Bureau was presented. This may be regarded as a sort of short-term prediction. The Hebei Province Seismological Bureau sent a field party for investigating seismological and geological situation to the Tangshan area on June 22. The investigation was finished on July 26.

Anomalies became more and more conspicuous towards the middle of July. Figure 8.1 shows the spike-like change in radon concentration observed at Langfang about 130 km distant from the epicenter. Figure 8.2 also shows the precursory change in earth resistivity observed at Changli about 80 km distant from the epicenter. Furthermore, outstanding macroscopic anomalies in underground water as mentioned in Subsection 4.10.1 began to be reported about 10 days before the earthquake. Anomalies in animal behavior were also so remarkable that the number of reports amounts to 2,003

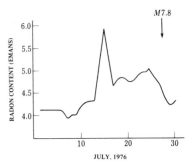

FIG. 8.1. Change in radon concentration observed at Langfang about 130 km distant from the epicenter prior to the Tangshan earthquake (WAKITA, 1978a).

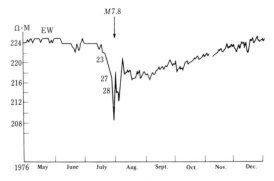

FIG. 8.2. Daily resistivity variation at Changli about 80 km distant from the epicenter prior to the Tangshan earthquake (NORITOMI, 1978a, b).

(OIKE. 1978a). As can be seen in Fig. 4.43, 80% of the reports of that type have a precursor time of one day or shorter.

In spite of such remarkable anomalies, no imminent prediction could be issued before the actual occurrence of the Tangshan earthquake. The following points may be regarded as the reasons of the failure. i) Most macroscopic anomalies appeared rather late than those for the Haicheng earthquake. ii) No conspicuous foreshocks were observed unlike the case of the Haicheng earthquake. iii) Appearance mode of precursory anomalies was so complicated that no anomaly was observed adjacent to a point where an outstanding anomaly was observed. For instance, many pigeons, which could not catch any precursory anomaly, died by collape of buildings in Tangshan. iv) Sympathizers of the gang of four did not pay much attention to earthquake information.

The disaster in Tangshan brought the dreadful result of failure in imminent prediction into relief. Even if a false prediction should be issued and people were evacuated as a result, no casualty would take place although economic activity in association with social confusion would be inevitable. As it is most important to save human lives, no failure in imminent prediction should be made even if false prediction could be made by chance.

In February, 1981, the author could after all visit Tangshan and see some of the earthquake ruins preserved by the order of local government. It would take some more years to reconstruct the city which was so terribly damaged.

8.4 Songpan-Pingwu Earthquakes

It had been anticipated in 1975–76 that a destructive earthquake would occur in an area about 200 km north-northwest of Chengdu, Sichuan Province at almost the same period as that of the Tangshan earthquake. Many large earthquakes had occurred along the Longmen Shan fault system as shown in Fig. 4.45. Earthquakes of $M \geqq 3$ have ceased to occur since 1973 in the Songpan-Pingwu area, which may be regarded as a seismic gap of the second kind (see Subsection 4.4.3). On the other hand, earthquakes of magnitude 5 or thereabout occurring around the area tended to concentrate along the

TABLE 8.1. The predictions of the Songpan-Pingwu earthquakes as presented by the Pingwu
Seismological Observatory to the author's seismological mission.

Date and origin	Event
Nov., 1975: Conference held by the Sichuan Province Seismological Bureau	A $M \geq 6$ earthquake might occur in the Maowen-Songpan area in the earlier half of the year 1976.
Jan., 1976: National Conference on Earthquake Prediction	The above view was approved. The area was designated to one of the areas of accentuated observation.
Jan., 1976	The Pingwu Seismological Observatory was established.
June 14, 1976: Earthquake Report No. 2 by Sichuan Province Seismological Bureau	Possibility of a $M \approx 6$ earthquake occurring in the Maowen, Beichuan and Kangding area within one to two months was pointed out.
June, 1976: Conclusion at a conference held by SSB and the Sichuan Province Seismological Bureau	Possibility of a $M \approx 7$ earthquake occurring in the middle southern portion of the Longmen Shan fault system within one to two months was pointed out.
July 30, 1976: Conclusion at the 1st Conference on Earthquake Prediction in the Pingwu Prefecture	Possibility of a destructive earthquake occurring in the Songpan-Pingwu area in the northwest part of the prefecture in the immediate future was recognized.
Aug. 2 and 7, 1976: Earthquake Reports No. 5 and 6 by Sichuan Province Seismological Bureau	It was announced to the administrative organizations concerned that a $M \geq 6$ or $M \approx 7$, in the worst case, earthquake seems likely to occur in the Maowen, Heishui, Kangding and Luding area along the Longmen Shan fault at around Aug. 13–17 and 23.
Aug. 9, 1976: Conclusion at the 2nd Conference on Earthquake Prediction in the Pingwu Prefecture	It was stressed that the possibility of a large earthquake occurring in the Songpan-Pingwu area is getting high, so that special attention should be paid to the time range put forward by the Sichuan Province Seismological Bureau.
Aug. 11, 1976: Determination by the Anti-earthquake Committee of the Pingwu Prefecture	Thirteen items for anti-earthquake arrangements were set out and conveyed to administrative units of various levels.

TABLE 8.1 (continued)

Date and origin	Event
Early morning on Aug. 12, 1976: Order by the Province Revolutionary Committee for imminent earthquake warning and associated preparation.	The Province Revolutionary Committee ordered the Sichuan Province Seismological Bureau and the Headquarters for Earthquake Disaster Prevention to communicate by telephone the imminent warming to earthquake offices of various levels, seismological observatories, observation points, amateur observation spots and other districts such as Miyanyang, Aba and Wenjiang. This is actually an order for alert disposition.
Aug. 12–14, 1976: All Pingwu Prefecture	The alert disposition was completed, so that men, cattle and other materials were moved to safe places.
Aug. 16, 1976	An earthquake of magnitude 7.2 occurred at the border between Songpan and Pingwu at 22:06.
Aug. 17, 1976	A party for analysis of earthquake state was formed in the Pingwu Seismological Observatory.
Aug. 21, 1976: Emergent conference at Pingwu	The following information was issued in the evening on Aug. 21. The information reads "A $M \approx 6$ earthquake would occur between 21:00 on Aug. 21 and 07:00 on Aug. 22".
Aug. 22, 1976	An earthquake of magnitude 6.7 actually occurred at 05:49 in the Songpan-Pingwu area.
Aug. 23, 1976: Emergent conference at Pingwu	On the basis of an emergent conference at Pingwu, it was publicized at 00:00 on Aug. 23 that a $M = 7.2$ earthquake would occur to the south of the previous shocks within 24 hours.
Aug. 23, 1976	An earthquake of 7.2 occurred about 10 km south of the previous epicenters at 11:30 on Aug. 23.

fault system and approach to the Songpan-Pingwu area since 1975. The Sichuan Province Seismological Bureau held a conference on the status of earthquake in Chengdu in November, 1975. It was concluded at the conference that a $M \geq 6$ earthquake seemed likely to occur in the Songpan-Maowen area in the early half of 1976 as is shown in Table 8.1. The table was prepared by the Pingwu Seismological Observatory and used by Li Dan-yang when giving a lecture on the process of prediction of the Songpan-Pingwu earthquake to the author's mission in September, 1978.

The above view was supported by the National Conference on Earthquake Prediction in January, 1976. Accordingly, the area was designated an area of accentuated observation, so that construction of the Pingwu Seismological Observatory and other observation stations was advanced. The Sichuan Province Seismological Bureau advised the administrative authorities of the possibility of a $M \approx 6$ earthquake occurring in the Maowen- Beichuan and Kangding area within 1–2 months' time on June 14. On the basis of the known tendency of alternate occurrence of large earthquakes between Yunnan and Sichuan Provinces as shown in Fig. 2.3, the members of Sichuan Province Seismological Bureau believed that they would have the next one in the Sichuan Province because the Longling earthquakes ($M = 7.5$, 7.6) have occurred on May 29 in the Yunnan Province.

SSB and the Sichuan Province Seismological Bureau held an emergency conference on June 22 and concluded that an $M \geq 7$ earthquake might occur at the middle and southern portion of the Longmen Shan fault system within 1 to 2 months' time. The Province Revolutionary Committee transmitted the information to revolutionary committees of various levels in the province. This is actually a short-term prediction of earthquake occurrence.

In order to obtain data for an imminent prediction, the Sichuan Province Seismological Bureau set up observatories at Heishui, Li Xian, Jiangyou, An Xian, and Shifang, and intensified gravity, magnetic, and levelling surveys. Amateur observation points, which were only 280 in number in 1975, have increased to 4,800. In spite of the fact that SSB and other organizations were busy beause of the medium- and short-term predictions in the Tangshan area, a substantial number of staff members and observation instruments were sent to the earthquake-threatened area in question from 13 provinces, cities, and autonomous districts.

As already stated in Subsection 4.10.1, anomalies of underground water and small animals started to be observed in an area of the Longmen Shan fault system to the south of Chengdu in June, 1976. Towards the middle of July, appearance of fire balls were reported in the Mianzhu area.

It is said that the following points were pointed out at the conference on June 22: i) Judging from the fact that expansion of the area in which anomalies were observed and that the time span, during which anomalies occurred, was extending, the expected earthquake magnitude seemed to amount to 6 or 7 according to past experience. ii) As few anomalies of large animals were observed and those of earth-currents, radon concentration and geomagnetic field not so conspicuous, the expected earthquake would occur towards the end of August. It is hard for the author, however, to reach such a conclusion from the reasons pointed out by Chinese colleagues.

The people in Chengdu and its surroundings panicked some time in July because of psychological fear of earthquake and the slander by sympathizers of the gang of four. It was thoroughly publicized, however, that the epicenter would be far from the Chengdu area, so that the confusion could be allayed.

A conference held in Pingwu on July 30 concluded that a large earthquake would occur in the Songpan-Pingwu area.

On August 2 and 7, the Sichuan Province Seismological Bureau issued Earthquake Reports Nos. 5 and 6 and informed the administrative organizations of the province of the possibility that a $M \geq 6$ or $M \approx 7$ earthquake in the worst case would occur in the Maowen, Beichuan, Kangling and Luding area on August 13–17 and around 23.

Towards the beginning of August, anomalies of geomagnetic field (Figs. 4.24 and 4.25), resistivity (Figs. 4.24 and 4.34), radon concentration (Fig. 4.41), underground water (Figs. 4.46 and 4.47), fire ball and so on became conspicuous as frequently referred to in Chapter 4. Those anomalies tended to concentrate to the Songpan-Pingwu area or area 3 in Fig. 4.45. The number of reports on macroscopic anomaly increased remarkably from the beginning of August as can be seen in Fig. 8.3 (Unpublished, Sichuan Province Seismological Bureau, 1978; OIKE, 1978a, b, c). The total number of reports amounted to 1,269 among which anomalies of underground or well water were 395 in numbers.

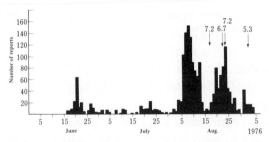

FIG. 8.3. Daily number of abnormalities in animal behavior and ground-water reported by the people before and after the Songpan-Pingwu earthquakes (Unpublished, Sichuan Province Seismological Bureau, 1978; OIKE, 1978a, b, c).

It was stressed at the 2nd conference in Pingwu on August 9 that much attention should be paid to the expected occurrence time of earthquakes in the Songpan-Pingwu area.

The Province Revolutionary Committee ordered the Seismological Bureau and the Anti-earthquake Headquarters that they should telephone all the organizations of various levels related to earthquake problems to take imminent precaution against the coming earthquake. This was actually an imminent warning. All the necessary arrangements for alert were completed on August 12–14, so that men, cattle, and important commodities were moved to safe places.

A violent earthquake of magnitude 7.2 occurred at last at 22 : 06 on August 16 at the border between Songpan and Pingwu Prefectures. The earthquake killed 21 and injured 360 people in the Pingwu Prefecture. The deaths were mostly due to land slides. Brick buildings were terribly damaged as can be seen in Fig. 8.4 although wooden houses were only slightly damaged. Most local inhabitants had slept in shelters as shown in Fig. 8.5 from July.

A party for analysing the state of seismic activity was formed at the Pingwu Seismological Observatory on August 17. The party consisted of all the 4 members of the observatory, 5 members of the Sichuan Province Seismological Bureau and a few members of seismological bureaus other than Sichuan Province, such as Nei Monggol.

An Earthquake Emergency Committee for the Pingwu Prefecture was also formed with the above party members, the Director of Earthquake Office and a few representatives (teachers, peasants,

FIG. 8.4. A brick building destroyed by the Songpan-Pingwu earthquakes at Shuijing located close to the epicenter.

FIG. 8.5. A wooden house, that was damaged only slightly, next to the brick building in Fig. 8.6 and a shelter where the people stay in case of an earthquake warning.

workers, and so on) from amateur observation points. A Committee meeting held in the evening of August 21 pointed out the possibility of a $M \approx 6$ earthquake occurring during a period between 21:00 on that day and 07:00 in the next morning. An earthquake of magnitude 6.7 actually occurred at 05:49 on August 22 in the Songpan-Pingwu area.

As a result of another Committee meeting held on August 22, it was disclosed at 00:00 on August 23 that a magnitude 7.2 shock seemed likely to occur in an area a little south of the former two epicentral areas within 24 hours. At 11:30 on August 23, an earthquake of magnitude 7.2 occurred as foretold at an area about 10 km south of the last two earthquake areas.

It is really surpising and admirable that the Chinese colleagues succeeded in imminent forecasting of the three successive earthquakes. The point that imminent warnings were put forward separately for each earthquake sounds almost incredible. It appears that the spike-like change in radon concentration at Guzan, about 320 km distant from the epicenters, was usefull for the estimate of occurrence time. As can be seen in Fig. 4.41, spike-like changes were observed both prior to the Luhuo ($M = 7.9$, 1973) and Mabian ($M = 5.5$, 1973) earthquakes each having an epicentral distance of 200 km. The precursor time was about 1 week for both cases. A sudden increase in the radon concentration by 70% was observed on August 10.

No outstanding foreshocks were observed for the Songpan-Pingwu earthquakes. However, it became obvious that the number of reports on macroscopic anomaly increases as the time of an earthquake approaches and decreases after the earthquake. When no decrease is observed, it is surmised that another large earthquake might follow successively.

When the author and his mission listened to the lecture given by Li Dan-yang at the Pingwu Seismological Observatory, the author raised a pertinent question about how the Chinese colleagues determined the time window for the dangerous period of earthquake occurrence. For instance, how the time window from 21 : 00 hours on August 21 to 07 : 00 hours on the next morining could be decided? The Chinese colleagues answered that there were differences in the estimate of time window between members of the Earthquake Emergency Committee for the Pingwu Prefecture, that no existing formula for estimating such a time window is available and that the time range is not so strict that the 07 : 00 hours may be replaced by 08 : 00 hours. The important point was, according to Chinese colleagues, that people should be evacuated during that night. In spite of these explanations, the author still feels that he does not fully understand the Chinese approach to imminent predictions which were surprisingly accurate in the result.

It is true that both instrumental and macroscopic anomalies become remarkable, that the number of reports increases enormously and that the anomalies tend to concentrate in an area which later becomes the epicentral area. As a result, according to Chinese colleagues, they feel ever-growing tension. At last, the tension increases to the highest degree, so that they are inclined to believe that an earthquake would certainly occur in the immediate future ranging 1–2 days. Such a judgment is called by Chinese colleagues a synthesized judgment. It appears to the author that the Chinese way of prediction is not quite objective. However, he admires with a great surprise the successes which Chinese colleagues accomplished.

It is doubtful that the Chinese way of approach can be applicable as it is to predictions in Japan. No such clear-cut appearance of macroscopic anomalies has ever been reported for destructive earthquakes in Japan in recent years. Probably, this is partly due to the fact that natural and man-made noise is high in Japan and partly due to the complicated crustal structure.

In spite of the frequent visits of U.S., Canadian and Japanese seismological missions to China, it cannot be said that the reason and background of the Chinese success in earthquake prediction became clear. The only way to fully investigate their success is to carry out internationally cooperative observation in the interior of the Chinese continent under the participation of field parties from one or more nations.

8.5   Izu-Oshima Kinkai Earthquake

Much of the land deformation, seismic activity, geomagnetic, and geoelectric changes, geochemical changes and the like preceding the Izu-Oshima Kinkai earthquake ($M = 7.0$, 1978) has been written in Chapter 4. As the earthquake occurred fairly close to the dense observation network for the feared Tokai earthquake as shown in Fig. 6.4, it was discovered that a number of precursory effects had been observed by detailed examination after the shock although it cannot be said that the earthquake was predicted.

It is the main aim of this section to summarize the precursory effects on the basis of the author's summary (RIKITAKE, 1981a). First of all, it should be emphasized that JMA issued earthquake information such as shown in Table 8.2 about 1.5 hours prior to the main shock as already stated in Subsection 4.4.1. The anomalous change recorded by a volume strainmeter at the southern extremity of the Izu

TABLE 8.2. The JMA earthquake information related to swarm earthquakes near the Izu-Oshima Island.

Time and date of issuance	Information
10:50 Jan. 14, 1978	Earthquakes began to occur near the Izu-Oshima Island since 20:38 on Jan. 13. The earthquake frequency has been so increasing that shocks are occurring almost continuously since 08:12 this morning.

The area is famous for earthquake swarms that occur almost every year. For instance, a swarm activity with 30 felt earthquakes took place there during Oct. 9–Dec. 11 last year. The swarm activity seems to be of fairly large scale comparable to that in December, 1964 when small damage was recorded.

It appears that a moderately large earthquake, that gives rise to slight damage, may occur shortly.

Jan. 13	08:12	Intensity III at Izu-Oshima Island
14	08:12	III
	09:06	III
	33	III
	36	IV
	38	III
	45	IV
	47	IV
	55	IV
	10:00	III
	08	III
	44	III
	46	III

Peninsula as shown in Fig. 4.10 along with the foreshock activity as shown in Fig. 4.12 played an important role in issuing the earthquake information.

The information by JMA was of course received by local authorities although some of the town and village offices did not pay much attention to the information because it was a few minutes before noon on Saturday. People working in those offices were just about leaving. Nobody, including many seismologists, expected such a large earthquake of magnitude 7.0 to occur there. Even the author thought that the maximum magnitude of an earthquake occurring there would not exceed 6.

TABLE 8.3. Geophysical and geochemical anomalies precursory to the Izu-Oshima Kinkai earthquake (RIKITAKE, 1981a).

Discipline	Precursor time (day)	Epicentral distance (km)	Observer	Observation point
Foreshocks	2		JMA	
Enormous increase in number of foreshocks	0.183		JMA	
Sudden decrease in foreshock activity	0.135		JMA	
Change in volume strain	42.5	32	JMA	Irozaki
Reversal of the above change	4	32	JMA	Irozaki
Change in volume strain	34.5	40	JMA	Ajiro
Change in geomagnetic field	64.5	38	JMA	Matsuzaki
Change in amplitude of short-period geomagnetic variation	69.5	30	ERI	Nakaizu
Change in earth potential	64.5	30	ERI	Nakaizu
Reversal of the above change	17	30	ERI	Nakaizu
Decrease in radon content	65.5	30	GI*	Nakaizu
Increase in radon content	7	30	GI	Nakaizu
Rise of water level	288.5	35	ERI	Funabara
Fall of water level	29.5	35	ERI	Funabara

* GI means Geophysical Institute, University of Tokyo.

It should be appreciated, however, that JMA issued a sort of earthquake information or warning. If a prediction council with the support of a well-equipped observation array had exsisted, it would not be unreasonable to think that much more accurate prediction information could be issued.

In Table 8.3 are reproduced the precursory effects summarized by RIKITAKE (1981a). Although the amount of data numbers only 14 in total, a histogram of logarithmic precursor time T measured in units of days is shown in Fig. 8.6. It is apparent, therefore, that the peak of the histogram of precursor time occurs at scores of days in the present case. As the frequency of precursor time becomes maximum at around 10 days or thereabout as can be seen in Fig. 4.55 for the world-averaged data, it may be that the mean precursor time for a particular earthquake can deviate from the mean value. This point should be important for an actual prediction.

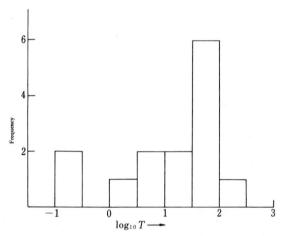

FIG. 8.6.   The histogram of logarithmic precursor time $\log_{10} T$, where T is measured in units of days, in the case of the Izu-Oshima Kinkai earthquake.

The peak of the histogram of precursor time of animal anomaly occurs at around 0.6 days as can be seen in Fig. 4.51. The precursor time for animal anomaly does not coincide with that of geophysical and geochemical precursors.

A number of Japanese seismologists feel that this earthquake could have been predicted if a well-organized array of multi-discipline observations had been in operation on the on-line real-time basis.

CHAPTER 9

CLASSIFICATION OF EARTHQUAKE
PREDICTION INFORMATION

9.1   Earthquake Prediction Information of Practical Use

As can be seen in the last chapter, short-term or imminent earthquake prediction information or warning so far issued to the public may typically read as follows; i) The probability of a slightly damaging earthquake occurring somewhere around so and so city, town or village within a few months is high. ii) The possibility of an earthquake having a certain magnitude occurring during a period between so and so o'clock on a certian day and so and so o'clock on the following day is high··· etc.

In order to express the risk of earthquake occurrence, we could use many words such as "probable", "possible", "in danger of", "feared", and so on. They could also be used in combination with adverbs such as "highly", "extremely", "fairly", "considerably", and so on. Logically speaking, these expressions are so vague, or inaccurate in other words, that subjective judgment about imminency may possibly be introduced by those who receive the information. It is very much so especially in the case of Japanese languages which are full of highly literary, but ambiguous, vocabularies.

On the other hand, earthquake information issuable to the public cannot use too scientific expressions. If the authorities should issue an earthquake warning that reads "It can be said that an earthquake will occur within so and so days *with a confidence level of so and so per cent"*. No ordinary citizens can understand the real meaning of the announcement. Moreover, even the officials who are responsible for preventing earthquake disaster would not know how to react against the warning stated in such unfamiliar words.

It is therefore requested by the public for earthquake prediction information to be issued, in such a way that administrators at various levels can react at once. When the Prediction Council for the Tokai area was formed, some discussion was made about the classification of earthquake prediction information of practical use. What follows is mostly due to RIKITAKE (1977b, 1978c) which was written for the purpose of forming the basis of such discussion.

First of all, the Council was asked by the government officials who are responsible for disaster prevention to classify earthquake prediction information into three classes according to time windows covering the dangerous period during which a high probability of earthquake occurrence is expected. The classification is as follows;

A: An earthquake would occur within several hours' time.

B: An earthquake would occur within a few days' time.

C: An eqrthquake would occur within a period that is not exactly known, but not very long.

Reliability of prediction information should also be rated in such a fashion as indicated below;

X: Highly reliable.

Y: Slightly less reliable.

Z: Possible.

It is therefore planned to present eqrthquake prediction information with a combination of (A, B, C) and (X, Y, Z). Administrative offices of various levels should in advance provide countermeasures for all the combinations. Should prediction information ranked at A or B with rating X be issued to the Tokai area, for example, action to stop the world-famous "Shinkansen" (bullet train) running through the area would be taken. Fire-brigades, police, self-defense forces, hospitals, and other governmental agencies would immediately stand by, too. If the rating is Y, the bullet trains may be operated at a maximum speed of 70 km/hr instead of their routine maximum speed of 210 km/hr.

The above emergency control of the bullet train is only one

example of governmental action relevant to earthquake prediction. Many other necessary arrangements such as evacuation of local people, closing of highways, suspension of nuclear power generation and the like would have to be made according to the combination of expected time window and its reliability rating.

Being a member of the Prediction Council, the author thinks that this request by governmental officials is quite right. If the prediction is purely academic as usually mentioned by means of confidence limit or something of that kind, no proper action would be able to be taken by government officials. The prediction should clearly specify the danger potential in such a fashion that practicable actions to prevent earthquake disasters effectively can be immediately taken. Consequently, we see that the situation resembles an assessment of a war council which is forced to decide to do something without having sufficient data to go on.

Although it is no easy matter to respond to such an administrative request at the present level of earthquake prediction study, the author (RIKITAKE, 1977b, 1978c) proposed a tentative approach to specifying the time window and its reliability as an earthquake prediction on the basis of analyses of earthquake precursor data so far collected.

9.2 Statistical Bases for Estimating Time Windows for Dangerous Periods

Although the main concern in this chapter is imminent prediction, it is natural to assume that the probability of an earthquake occurring in an area in question has been estimated in advance to be high on the basis of statistical and long-term (see Subsections 5.4.1 and 5.4.2) and medium- and short-term (see Subsection 5.4.3) predictions.

For the purpose of imminent prediction, we have to rely on precursors having a precursor time ranging from a few hours to a few days. Let us then make a Weibull distribution analysis of the frequency distribution of precursors of the third kind as shown in Fig. 4.55. According to RIKITAKE (1978c), the mean value of $\log_{10} T$ (T is the precursor time measured in units of days) amounts to 0.66 ($T=4.6$

days) with a standard deviation amounting to 0.95. Making use of two parameters of Weibull distribution best fitting the histogram in Fig. 4.55, it is possible to estimate the cumulative probability of an earthquake to occur. The probability increases as time goes on as can be seen in Fig. 9.1.

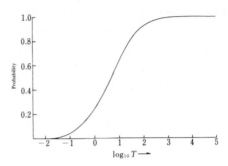

FIG. 9.1. Changes in cumulative probability of an earthquake occurring with increase in logarithmic time $\log_{10} T$, where T is measured in units of days, when a precursor of the third kind is observed at $\log_{10} T = -3$.

Let us define, rather arbitrarily, the dangerous time window by the period for which the cumulative probability takes on values ranging from 0.3 to 0.8. If an earthquake does not occur, even if the estimated probability exceeds 0.8, we assume that the signal was false. In that case the time window for a precursor of the third kind becomes 1.4–28.2 days.

A similar estimate for precursors of the second kind, which has a peak of frequency histogram at around $\log_{10} T = -1$ as can also be seen in Fig. 4.55, leads us to a mean precursor time and its standard deviation respectively amounting to 2.4 and 2.3 hours. The dangerous time window is then estimated as 1.0–4.0 hours.

If we further make use of the statistics of anomalous animal behavior although we have to admit that the quality of data would not be high, the mean value of $\log_{10} T$ is estimated as -0.4 ($T = 0.4$ day) along with the standard deviation amounting to 1.0 (RIKITAKE, 1978b). In that case the dangerous time window is estimated as 0.13–4.0 days.

Judging from the time windows obtained in the above, it may be said that classification A, B, and C can be provided, to a rough degree of approximation, by precursors of the second kind, of animal behavior and of the third kind, respectively.

9.3 Reliability of Prediction Information

In order to classify earthquake prediction information into A, B, and C stages as mentioned in the last section, it is clear that observations of precursors of various disciplines as listed in Table 4.1 are indispensable. The abbreviations shown in Table 4.1 are also used in Table 9.1 along with a that represents animal anomaly.

In Table 9.1 are shown the types of precursor, the dangerous time windows estimated from those precursors and the disciplines of precursor that may be useful for predictions of respective types.

One of the difficulties in classifying precursors is certainly the point that, even if a signal is detected, we do not generally know to which class of precursor it belongs with a few exceptional cases such as a signal caught by the Yamazaki variometer (see Subsection 4.6.5; YAMAZAKI, 1975; RIKITAKE and YAMAZAKI, 1976, 1977). The variometer usually records a precursory resistivity change of the second kind although some change of longer time-span could be observed with a careful processing (RIKITAKE and YAMAZAKI, 1979).

RIKITAKE (1975c) has shown, however, that, when a precursory land deformation is observed, the probability for the signal belonging to a precursor of the first, second and third kinds is estimated as 57, 11, and 32%, respectively. It may sometimes be possible to judge to which type of precursors a signal belongs on the basis of space-time pattern of occurrence of the signal. Although there may be some cases for which we can infer the class of precursory signal, it is in general difficult to make a distinction mechanically.

There is no established method of specifying the reliability of prediction information. RIKITAKE (1977b, 1978c) tentatively proposed that such a rating should be made by taking into consideration the percentage of disciplines for which anomalous signals are detected and that of the number of observation stations at which similar signals are detected. These percentages are denoted by λ and μ, respectively. The ratings X, Y, and Z as mentioned in Section 9.1, may for instance be specified by combinations of λ and μ as shown in

TABLE 9.1. Practical classification of earthquake prediction information*.

Precursor type	Imminent			Medium- and short-term	Statistical and long-term
	A Precursor of the second kind	B Animal anomaly	C Precursor of third kind	Precursor of the first kind	Crustal strain and historical document
Time window	1–4 hours	3 hours–4 days	1.5–28 days	Several years	Scores of years
Precursor discipline	$l, t, f, m, g, e, r, u, a$	a	l, t, f, m, g, e, u	$l, t, f, b, m, s, c, v, w, g, e, r, i, u, o$	l, active fault, historical record
Reliability X	$\lambda \geq 75,\ \mu \geq 50$		$\lambda \geq 75,\ \mu \geq 50$	Magnitude is estimated from spatial extent of anomaly, and then precursor time is estimated from magnitude	Probability of earthquake occurrence is estimated for each year
Y	$75 > \lambda \geq 25,\ \mu \geq 25$ $\lambda \geq 75,\ 50 > \mu \geq 25$		$75 > \lambda \geq 25,\ \mu \geq 25$ $\lambda \geq 75,\ 50 > \mu \geq 25$		
Z	$\lambda > 25,\ 25 \geq \mu$ $25 \geq \lambda,\ \mu \geq 25$	Many reports on anomalous animal behavior in a certain area	$\lambda > 25,\ 25 \geq \mu$ $25 \geq \lambda,\ \mu \geq 25$		

* See text and Table 4.1 for notations; λ and μ are in percentage.

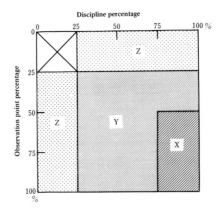

FIG. 9.2.   Rating of reliabilities *X*, *Y*, and *Z* by means of combinations of discipline and observation point percentages.

Fig. 9.2 and Table 9.1.

It is doubtful in Japan, where we have no systematic observation of anomalous animal behavior at present, that discipline *a* or animal behavior can play an important role in earthquake prediction although it is listed in Table 9.1. According to information the author gained, in China, it cannot be ruled out that the rate of increase of reports on macroscopic anomalies including animal anomaly can be used for specifying time windows expressed by A, B, and C at least in China.

The classification of earthquake prediction information as presented here is based on the author's personal view. It is clear that such a classification has been based on data not fully established, so that the classification may not at once be applied to actual prediction as it is. However, the author hopes that a classification of earthquake prediction information as proposed here could be achieved in the long run of earthquake prediction study in association with the increase in knowledge on earthquake precursors.

At the moment, most countermeasures that will be taken by the national and local governments and other agencies in Japan are assuming an emergency state expressed by **AX**. The author thinks that this is quite natural. If the prediction information is going to be CZ, a fairly indefinite one, the officials responsible for disaster prevention do not know how to react. In a capitalistic country like Japan, it is almost impossible to take an action which suppresses economic activity over a long period of time.

CHAPTER 10

FORM OF EARTHQUAKE WARNING

10.1 What is an Earthquake Warning?

No definite and well-documented warning for a coming earthquake has ever been issued to the public from accountable authorities except the Chinese instances. Accordingly, no established form of earthquake warning has as yet been agreed, especially in the western countries.

What should be an earthquake warning? The question may be partly answered by the following which was taken from NATIONAL ACADEMY of SCIENCES (1975):

"It is useful in developing public policy to make a clear distinction between *prediction* and *warning*. *Prediction* is a statement indicating that an earthquake of a specified magnitude will probably occur at a specified location and time, based on scientific analysis of observed facts. A prediction is strictly information; it says nothing about how people should respond and takes no account of the consequences that may follow from the issuance of the prediction. Issuance and assessment of predictions are strictly technical grounds.

A *warning*, on the other hand, is a declaration that normal life routines should be revised for a time. Warnings are issued because of a judgment that public welfare will be served thereby. Warnings are normally based on predictions or other types of technical information, but not all predictions will be followed by warnings. Issuance and assessment of warnings are peculiarly the responsibility of public officials acting in the interests of the people they represent".

In other words, it is clearly pointed out that an earthquake warning is to be handled by administrative authorities who may add an administrative judgment to earthquake prediction information, whereas earthquake prediction is a purely scientific and/or technical matter to be handled by scientists.

The author gives whole-hearted support to the above view because he believes that scientists should make earthquake prediction without paying attention to social effects that may arise as a result of prediction. If scientists involved are so concerned that economic depression might hit the area under consideration, that local people might meet difficulties, and so on, they may be influenced from making a fair, objective judgment.

However, the situation is different for administrators. They may modify the earthquake prediction information after administrative consideration, if necessary. They may also choose the best timing for issuance of a warning. At any rate it is absolutely important that an earthquake warning should be issued by officials responsible for administration and not by scientists.

When the Prediction Council for the Tokai area was formed, there was much debate about who makes the earthquake prediction information public. Some government officials insisted that scientists should do so because they know the situation best. After serious discussions, however, it was finally decided that the release of information or warning is to be made by the Japanese government which would take administrative conditions into consideration before doing so.

In an actual emergency case, however, there is no guarantee that things proceed as mentioned in the above paragraphs. When news that the Prediction Council is to be summoned is distributed, many citizens would be panicked. News reporters would endeavor to obtain comments in advance from the Council members who can hardly avoid journalists. The mass media is so developed in Japan that Council members could easily be caught by pressmen, wherever they go. It is unfortunate that what Council members say could be modified and distorted when reported by the news media.

It has been customary that the CCEP President has a press conference after a CCEP meeting. The President usually explains what has been discussed at the meeting in relation to data that might have something to do with long-range earthquake prediction. CCEP being only an advisory organ to GSI's Director, the President would not talk about earthquake prediction information because he has no official right or responsibility for doing so as mentioned in Subsection 6.1.4. He might say, however, that an anomalous uplift has been found in so and so area. Although he is just going to point out only the fact of anomalous uplift, the pressmen would like to have details and additional points. For instance, they might further ask "What magnitude would it be, should the anomaly be connected to an earthquake?" The answer to such a hypothetical question would then become an important news sometimes causing a panic-like confusion in the locale. It is sometimes difficult for seismologists to be very frank because of the situation mentioned above.

In conclusion, it is no easy matter to establish a suitable form of earthquake warning or prediction information within a short period of time because it is deeply concerned with social, socio-psychological, economic, legal and political problems. It is very important, nevertheless, to establish a guide line for issuing an earthquake warning in order to prevent earthquake disaster, on the basis of earthquake prediction. Although it cannot be avoided to do this in a fashion of trial and error in the beginning, however, it is hoped to reach a reasonably successful level within the foreseeable future.

10.2 Earthquake Warning in Japan

No information officially called an "earthquake warning" has ever been issued to the public in Japan. When swarm activity has already been experienced, JMA is in a position to offer "earthquake information" that is concerned with the present status and expected future progress of the activity to the public in conformity with the Meteorological Service Law. The earthquake information for the Matsushiro and Izu-Oshima Kinkai earthquakes as mentioned re-spectively in Sections 8.1 and 8.5 was put forward on JMA's responsibility although it was practically an earthquake warning. In association with the enactment of the Large-scale Earthquake Countermeasures Act, it was officially decided that the Prime Minister issues an "earthquake warnings statement" in case an imminent prediction is put forward by the Prediction Council for Areas under Intensified Measures against Earthquake Disaster.

It is thus made clear that an earthquake warning should be issued on the full responsibility of the Prime Minister, who represents the Japanese Government as far as the large-scale earthquake specified before is concerned.

It is the aim of this section to describe how and why the earthquake warning policy in the Tokai area has been established in recent years. Although the effort towards establishing ways and means for preventing earthquake disaster finally resulted in the enactment of the Large-scale Earthquake Countermeasures Act, bureaucratic red tape must be cut in order to complete such counter-measures not yet undertaken.

10.2.1 The feared Tokai earthquake and the proposed system for issuance of earthquake warning

It has often been mentioned in Subsections 1.1.1, 3.1.2, 6.1.5, 6.1.6, and elsewhere in the foregoing parts of this book, that enormous crustal strain has been accumulated in the Tokai area, that the area was designated an "area of intensified observation" by CCEP, that the Prediction Council for the Tokai area was set up, that the special legislation was made aiming at preventing damage that would be caused by the expected earthquake, if occurred, and that the Tokai and neighboring area was designated an "area under intensified measures against earthquake disaster". It is the author's intention in this subsection, to summarize such developments in relation to earthquake prediction/warning work in the Tokai area.

(1) Impact of the Ishibashi hypothesis

It has been commonly recognized by most seismologists in Japan that much strain had been accumulating in the earth's crust of the Tokai area, where no great earthquake has occurred for more than 120 years, since, in fact, the Ansei Tokai earthquake ($M = 8.4$, 1854). The mean return period of recurrence of great earthquakes there is estimated as 120 years or thereabout. As can be seen in Table 1.2, CCEP, which paid attention to the above fact, designated the Tokai area an "area of intensified observation" on Feb. 28, 1974. The epicenter of the feared quake used to be supposed somewhere in the Enshunada Sea, off the Pacific coast of Shizuoka Prefecture.

K. Ishibashi, who paid attention to the land uplift which was recorded in historical documents found anew in relation to the Ansei Tokai earthquake, suggested that the epicentral area of the quake must have been extended into the Suruga Bay in contrast to the previous supposition so far accepted and that the coming earthquake could have a similar epicentral area. Ishibashi, who attended a CCEP meeting as an observer reported his view there, while he read a paper of a similar kind at the 1976 fall meeting of the Seismological Society of Japan. It is the author's understanding that the hypothesis put forward by Ishibashi did not include any discussion about the occurrence time of the expected earthquake. Nevertheless, Ishibashi's view was modified in a fashion that influenced local people to think that a great earthquake might imminently occur somewhere in the Suruga Bay. As the Shizuoka Prefecture becomes the target of such a great earthquake that would occur immediately underneath Shizuoka, Shimizu, Yaizu and other cities, social unrest was widely spread.

In order to deal with such an unusual state of affairs, the national and local governments and the seismological community took actions necessary for suppressing such social anxiety as listed in Table 10.1. These actions finally resulted in the enactment of the Large-scale Earthquake Countermeasures Act in 1978 and the designation of the area under intensified measures against earthquake disaster on August 7, 1979.

(2) Actions taken by Shizuoka Prefectural Office

When Ishibashi talked about his hypothetical theory, an earthquake swarm activity had been prevailing in the East Izu area of Shizuoka Prefecture and the Kawazu earthquake ($M = 5.5$) happened to occur there on August 18, 1976. Under the circumstances, citizens who lived in Shizuoka Prefecture tended to show concern that a great earthquake might hit the area. The Governor then announced on August 28 that he would start a working group which investigates

248

TABLE 10.1.   Scientific and administrative events related to the earthquake prediction of the Tokai earthquake (SCIENCE AND TECHNOLOGY AGENCY, 1978 and other sources).

Date	Event
1976	
Aug. 18	Kawazu, Izu Peninsula, earthquake ($M = 5.5$).
23	K. Ishibashi, a research associate at the Geophysical Institute, University of Tokyo, reported on a new view about the epicentral area of the expected large earthquake in the Tokai area at a regular meeting of CCEP.
Oct. 1	An anti-earthquake team consisting of 4 members headed by I. Koshii was formed in the Section of Fire Defense and Disaster Prevention, Shizuoka Prefectural Office.
4	T. Asada, a professor at the Geophysical Institute, University of Tokyo, testified the possibility of the Tokai earthquake at the Budget Committee of the House of Councillors.
8	Ishibashi's report at the fall meeting of the Seismological Society of Japan in Fukuoka City. Minister of Science and Technology announced possible establishment of an organ that deals with the suspected Tokai earthquake.
13	T. Hagiwara, the President of CCEP, testified at the Committee for Promotion of Science and Technology of the House of Representatives that the CCEP supports the view that a great earthquake would occur in the Suruga Bay area although no data was available for occurrence time.
21	Special Committee for Earthquake Prediction of the Geodetic Council discussed the structural organization of various observations in the Tokai area in order to cope with imminent prediction of the suspected earthquake.
1976	
Oct. 29	Headquarters for Earthquake Prediction Promotion was set up and attached to the Cabinet.
Nov. 4	Anti-earthquake Committee for Main Roads was formed in Shizuoka Prefectural Office.
12	Shizuoka Prefectural Office presented a report on the intensification of the earthquake prediction system and subsidy arrangement for anti-earthquake work to the Chairman's Meeting of the Ten Prefectural Assemblies.
25	Anti-earthquake Working Group was set up in the Shizuoka Prefectural Office.
29	CCEP released its unified view that the probability of occurrence of the said Tokai earthquake is high although no precursory effect has been observed.
Dec. 17	Geodetic Council proposed the establishment of the Prediction Council for the Tokai area.

possible intensification of anti-earthquake measures and possibility of earthquake prediction. On the same day, the Liberal Democratic Party of the prefecture put forth a view that a watching system for

TABLE 10.1 (continued)

Date	Event
1977	
Apr. 4	Headquarters for Earthquake Prediction Promotion decided to set up the Prediction Council for the Tokai area.
18	First meeting of the Prediction Council for the Tokai area.
Aug. 1	Anti-earthquake Section was set up in the Shizuoka Prefectural Office.
Oct. Dec.	Preparation for special legislation aiming at mitigating earthquake hazards on the basis of earthquake prediction information was made. Two tentative plans were drafted by S. Harada, a Dietman of the Liberal Democratic Party and All Japan Conference of Prefectural Governments.
1978	
Jan. 14	Izu-Oshima Kinkai earthquake ($M = 7.0$).
17	Prime Minister instructed at a Cabinet meeting to investigate special legislation for dealing with a large-scale earthquake.
Feb. 17	The outline of the Large-scale Earthquake Countermeasures Act was made public by the National Land Agency.
Apr. 5	The above act was presented to the Diet.
June 7	The above act passed the Diet.
Dec. 14	The above act was enacted.
1979	
Aug. 7	Shizuoka Prefecture and neighboring area was designated the Area under Intensified Measures against Earthquake Disaster by the Large-scale Earthquake Countermeasures Act. The Prediction Council for the Tokai area was reorganized as the Prediction Council for the Area under Intensified Measures against Earthquake Disaster.

earthquake prediction serving the prefecture itself should be established. Meanwhile the "Komei" Party of the prefecture asked the Governor to set up the Section for Earthquake Disaster Prevention, to intensify anti-earthquake measures, to commission experts for earthquake prediction, and to revise the system for dealing with emergency as proposed by the Section of Fire Defense and Disaster Prevention of the Prefectural Office.

The Governor started an Anti-earthquake Team in the Section of Fire Defense and Disaster Prevention on October 1. The team was later promoted to the Anti-earthquake Section. The Anti-earthquake Committee for Main Roads and the Anti-earthquake Working Group were successively set up on November 14 and 25, respectively. The latter consists of 9 subgroups; i.e. emergency commodities, assessment of earthquake and fire damage, measures for traffic control, measures of sea problem, measures for land-slides and floods, mass mobilization, water supply and evacuation.

(3) Actions taken by cities, towns and villages in Shizuoka Prefecture

It was naturally planned to intensify anti-earthquake measures at the level of city, town and village and to promote the spread of earthquake knowledge. A civil defense system has also been promoted.

(4) Action taken by the Japanese Government and the Diet

T. Asada, a professor at the Geophysical Institute, University of Tokyo, testified to the possibility of the Tokai earthquake at the Budget Committee of the House of Councilors on October 4, 1976.

In order to promote administrative business between ministries and agencies involved and to push overall and intentional policies in relation to earthquake prediction, the Headquarters for Earthquake Prediction Promotion, headed by the Minister of Science and Technology, was formed and attached to the Cabinet on October 29 as has already been stated in Subsection 6.1.4.

On November 29, CCEP released its official view that the probability of a large earthquake occurring in the Tokai area is high although no precursory effects, by which the occurrence time can be guessed, have been observed.

The ways and means for achieving imminent prediction of the feared earthquake were discussed at the Geodetic Council which proposed the 2nd revised proposal for the 3rd 5-year program on earthquake prediction to the Government on December 17, 1976. On the basis of the proposal the Headquarters for Earthquake Prediction Promotion decided to set up the Prediction Council for the Tokai area on April 4, 1977. The Council has officially started on April 18.

(5) Legislation of the Large-scale Earthquake Counter-measures Act

K. Yamamoto, the Governor of Shizuoka Prefecture, has proposed forming a unified system of earthquake prediction and disaster prevention. He asked the All Japan Conference of Prefectural Governors to set up a Special Committee for Anti-earthquake Measures on July, 1977. The committee approved the outline of a special legislation with respect to earthquake countermeasures on December 7, 1977. The outline emphasizes the following points;

   i)   Intensification and systematization of national earthquake prediction observation and issuance of earthquake warning.

   ii)   Legislation of duties of administrative organizations and citizens, emergency means to be taken by the prime minister, and emergency measures to be taken by local governments such as prefectures.

   iii)   Planning and promotion of special undertaking for disaster prevention.

The All Japan Conference of Prefectural Governors requested the national government to make a special legislation that involves the above outline which is the highlight of its draft proposal.

Preparation of such a legislation has also been made by the Fire Defense Board and other organizations of the Japanese Government. Debate has also been going on at the Special Committee for Anti-earthquake Measures of the Liberal Democratic Party resulting in a draft act called the Harada proposal. S. Harada, a member of the House of Representatives, worked hard on drafting the proposal.

In spite of the above movements longing for special legislation, no remarkable reaction of the Japanese Government took place until the occurrence of the Izu-Oshima Kinkai earthquake ($M = 7.0$, Jannuary 14, 1978). At the Cabinet meeting on January 17, the Prime Minister asked the Minister of National Land Agency to investigate the details of special legislation.

Actual work of drafting the law was conducted by the National Land Agency, which put stress on the draft by the All Japan Conference of Prefectural Governors and adjusted many views of various governmental organizations involved.

The outline of the "Large-scale Earthquake Countermeasures Act" was thus publicized on February 17, 1978 after painstaking work of debating and adjusting. The Act was presented to the Diet on April 5, passed the Diet on June 7 and enacted on December 14 as stated earlier in this book. As the details of the Act will be mentioned in Chapter 13, the author here points out that, according to the Act, an earthquake warning should be issued only by the Prime Minister. It has now become clear that an earthquake warning is issued on the responsibility of the Japanese Government.

10.2.2 Examples of judgment by the Prediction Council for the Area under Intensified Measures against Earthquake Disaster

As of November, 1979, no precursory effects were observed in the Tokai area which was designated an Area under Intensified Measures against Earthquake Disaster by the Large-scale Earthquake Countermeasures Act on August 7, 1979. Investigations undertaken by the Prediction Council for the Area under Intensified Measures against Earthquake Disaster will determine what kind of style must be adopted for the report on an imminent prediction which should be conveyed to the Prime Minister via JMA's Director General and finally to the public in association with the Earthquake Warnings Statement to be issued by the Prime Minister.

It is anticipated by the Council that much time cannot be available for discussing and investigating over the style of the prediction information report in case of emergency. If so, it will be helpful for the Council to draft the structure of such a report well beforehand. It is planned that the draft can be used in an actual case by just filling in the blank portions for times and locations.

When the Council expects the occurrence of the feared earthquake, one of the typical examples of report would be something as follows;

"Since around (*hour*) on (*day*), (*month*), (*name of instrument*) at (*location*), (*name of instrument*) at (*location*), · · ·have been simultaneously observing anomalous changes and (*small/moderately large*) earthquakes have been occurring (*so and so times an hour*) in the (*location*) area.

Judging from these phenomena, it is feared that a large-scale earthquake will occur in the epicentral area centering on (*location*) within 1) several hour's time or 2) a few days' time.

Should the earthquake occur, the intensities will be (*intensity*) at (*locality*), (*intensity*) at (*locality*), · · ·. It is also feared that tsunamis will hit the coasts of (*locality*), (*locality*), · · ·".

Only two time windows are chosen, so that either "within several hours' time" or "within a few days' time" are to be used in an actual case. It is thought that confusion could arise if we set out many time windows. It could happen that the dangerous time window should be elongated after the period of the first announcement. In such a case, extension of dangerous period, having a similar time window as before, may be announced.

A few more examples of similar standard reports are available, for cases in which the occurrence of a great earthquake is expected (although they are not quoted here for the sake of simplicity). These are slightly different from the above example in the description of precursory effects.

When the Council decides that the earthquake probability is low, a report something like the following will be presented;

"From approximately (*hour*) on (*day*), (*month*), (*name of instrument*) at (*location*), (*name of instrument*) at (*location*),···have been observing anomalous changes and (*small/moderately large*) earthquakes have been occurring (*so and so times an hour*) in the (*location*) area. After careful investigations, however, it is concluded that the probability of a large earthquake occurring there is low. However, citizens are requested to pay attention to earthquake information that follows".

At any rate, it is planned to draft standard reports as simply as the Council can, as cited above. The process through which the Council reached its judgment will be made public by the Chairman when he gives a detailed account of the situation to the public.

10.2.3 Progressive access to earthquake warning

As mentioned in the last subsection and in Table 10.1, CCEP understands that no precursory effects of the 1st, 2nd and 3rd kinds have been observed in the Tokai area. Such a situation seems likely to continue up to 1981. If we assume that precursory phenomena would appear before the expected Tokai earthquake in a fashion similar to the averaged tendency as presented in Section 4.11, it would take some time before the actual occurrence.

It is not unreasonable to suppose that we would have precursors of various kinds before the catastrophe, although the possibility that we would not detect any other precursors except an imminent one cannot be ruled out. In order to cope with such a case, the Prediction Council is always on the alert.

It is the author's view, however, that the earthquake prediction information would become accurate year by year because of possible detection of long- and medium-term precursors.

10.3 Earthquake Warning in the U.S.A.

10.3.1 How to issue an earthquake warning?—models of earthquake warning

Socioeconomic studies on the effect of earthquake prediction/warning commenced in the U.S.A. in 1975. It was widely believed at that time, that earthquake prediction would become a matter of actuality in the foreseeable future. Studies entitled "A technology assessment of earthquake prediction" by the Stanford Research Institute and "Socioeconomic and political consequences of earthquake prediction" by the University of Colorado were then started with the support of NSF.

Although the likely reactions of the public against an earthquake warning will be discussed in Chapter 12, it has gradually become clear what style an earthquake warning must take all through the above and associated studies. A conference on earthquake warning and public reaction against it was held in San Francisco on November 7, 1975 (U.S. GEOLOGICAL SURVEY, 1976). On that occasion the USGS's Director reported that an earthquake warning should be issued by the state and other local governments although USGS is responsible for earthquake prediction information as can be seen in Fig. 10.1.

In order that an earthquake warning/earthquake prediction information can be seriously taken, believed and made use of by the public, it is not enough that prediction is scientifically correct but is also seen and understood as a clear warning. WEISBECKER *et al.* (1977) pointed out that the following conditions must be fulfilled for practical earthquake prediction:

(1) The prediction must in fact be a warning: it must convey a sense of danger, not just neutral "scientific" information.

(2) The warning must be specific as to time, place, and intensity

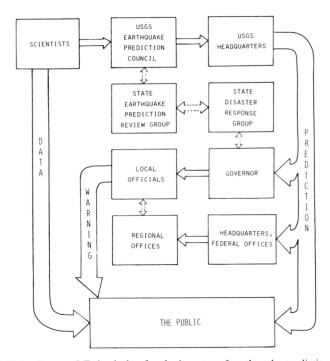

FIG. 10.1. Proposed Federal plan for the issuance of earthquake predictions and warnings. The plan provides for continual public release of scientific data but ensures that there are firm bases for an official prediction. In the plan, the USGS has the responsibility for issuing a prediction (statement that an earthquake will occur). Whereas State and local officials have the responsibility for issuing a warning (recommendation or order to take defensive action) (U. S. GEOLOGICAL SURVEY, 1976).

and must accurately identify the areas, people, and structures risk.

(3) The warning must contain prescriptions or at least strong suggestions for action.

(4) The warning must be disseminated through society's existing communication system, using both formal and informal networks of communication.

(5) The warning system as a whole should adhere to the "principle of redundancy"; that is, it should provide alternative independent sources of communication that are mutually confirming and consistent when cross-checked.

WEISBECKER *et al.* (1977) also suggested a typical form of

TABLE 10.2. Typical example of earthquake prediction information in California (WEISBECKER et al., 1977).

Element	Remark
Lead time (months)	6
Time window (weeks)	±3
Epicenter or region of fault ruptures	San Juan Bautista to Los Gatos along the San Andreas Fault
Magnitude (Richter)	7.0 to 7.2
Confidence that event will occur (percent)	85
Contingent effects	Possible 8.3 Richter magnitude along entire "locked" San Francisco Bay section of the San Andreas Fault (no confidence judgment possible)

earthquake prediction information relevant to the San Andreas fault as shown in Table 10.2. The author does not know how the time window and occurrence probability cited in the table can be estimated. It may in practice be difficult to estimate such quantities accurately.

10.3.2 Governor's decision

It appears to the author that the seismicity in the U.S.A. is so low that no immediate threatening of a great earthquake is acceptable as a matter of actuality by citizens even in California. Accordingly, preparatory countermeasures against an expected earthquake, such as that taken by the Governor and his government of Shizuoka Prefecture, Japan, have not found a similar response by the Governor of California . The only suggestion as to what the Governor of California would do when an imminent information of earthquake prediction reached him was provided by WEISBECKER et al. (1977) as a fictitious scenario.

According to WEISBECKER et al. (1977), the scenario was prepared in order to suggest several possible outcomes and to illustrate the difficulties faced by a decisionmaker in choosing to translate a set of scientific facts into a warning of a specific event.

TABLE 10.3. *Hypothetical* setting of short-term prediction information in California (WEISBECKER *et al.*, 1978).

Date	Event
Feb. 3, 1979	The USGS obtained some indications of surface-level changes at two stations 20 miles apart along the southern end of the locked section of the San Francisco Bay portion of the San Andreas Fault.
March 15, 1979	The 20-mile segment is instrumented with a fairly dense network of tiltmeters.
June 15, 1979	Using the installed instrumentation as well as regular and systematic measurements of resistivity, telluric currents, magnetic anomalies, radon emission from deep wells, water levels in deep wells, microquake measurements, V_P/V_S anomalies, and Vibraseis surveys, the earthquake premonitors are isolated to a 40-mile section of the San Andreas Fault between San Juan Bautista and Los Gatos. This is translated into a prediction that there is an 85 percent chance that an earthquake of magnitude 7 to 7.2 will take place on December 15, 1979, plus or minus 3 weeks.
June 30, 1979	After receiving confirmation from USGS headquarters in Washington, the USGS formally notifies the Governor of California that it would release this information in 24 hours, which gives him time to make a decision concerning a warning on these facts. They also notify the Governor that there is a real possibility that the predicted earthquake may trigger a great earthquake along the entire locked section of the San Andreas Fault.

(1)   Hypothetical setting of situation

It is assumed that USGS observed anomalous signals which are likely to be connected to an earthquake occurrence. The information has been officially reported to the Governor by USGS. The hypothetical situation is summarized in Table 10.3 in which the dates and information are all fictitious.

(2)   Difficulties in the Governor's decision

It was already stated in Section 6.2 that the Governor of California has his own California Earthquake Prediction Evaluation Council. According to the scenario, the experts of the panel approved the USGS work. Actually, some of the members of the panel have been joining in the work conducted by USGS.

The Governor called his panel of policy experts and asked their advice on action. He explained to the panel that he has the following alternatives possibly affecting his option:

i)   With respect to the prediction

The prediction may stand unaltered.

The prediction may be modified to represent a different set of facts.

The prediction may be made more precise with a short-term warning having a reasonable degree of confidence.

ii)   With respect to the earthquake

It may not happen at all.

It may happen as predicted.

It may happen outside the parameters of the prediction. It may trigger a 1906-type earthquake in the San Francisco Bay Area.

iii) With respect to his response to the present prediction

He can issue the present prediction for information only for local option action.

He can issue the present prediction for information only, with the understanding that we have a good chance of getting a short-term warning.

He can issue a warning that would initiate action based on the present information. Of course, this would be appropriately modified if and when we received further information.

He can do any of the above either for a 7 to 7.2 magnitude earthquake that would affect the area in which premonitors are found or for an 8.3 magnitude earthquake that would affect the entire San Francisco Bay Area, or he can select a combination of the above for both situations.

It is understood that each of these options is tightly connected to public risk. The effect of any option on different segments of society may differ considerably. The Governor feels that he is facing political risk. The optimum public interest may differ from the optimal private interests of his supporters.

Frankly speaking, he does not know what to do. To make matters worse he has only 24 hours to decide. The scenario ideally describes the Governor's annoyance. It is unlikely, however, that answers on decision-taking can be shown on the basis of a fictitious scenario.

Although it is instructive and good for understanding to explain the problem by means of fictitious scenario as presented above, the author feels that the persuasive power is a little lacking compared to actions the Governor and his government of Shizuoka Prefecture are at present implementing because the latter is actually facing-up to the menace of a great earthquake.

10.4 Earthquake Warning in China

The form of earthquake warning and the system for issuing a warning have been explained in detail in Sections 8.2, 8.3, and 8.4 respectively for Haicheng, Tangshan and Songpan-Pingwu earthquakes. Especially, it is clearly shown in Table 8.1 how earthquake warnings are put forward in China. In short, earthquake prediction is the job of SSB, Province Seismological Bureaus, seismological observatories and seismological observation spots, while earthquake prediction information is to be summarized by Earthquake Offices of various levels and earthquake warnings are issued by Revolutionary Committees of various levels after applying administrative consideration. Usually, no long-term prediction is issued to the public. Such a prediction is conveyed only to staff members of administrative offices of various levels.

It should be noted that earthquake prediction information is provided by scientists without paying attention to administrative matters and that earthquake warning is to be issued by administrators even in China just like in the U.S.A. and Japan despite the difference in the social organization.

COMMUNICATION OF EARTHQUAKE PREDICTION INFORMATION AND WARNING

11.1 Long-Term Earthquake Prediction Information

It has recently become common practice in Japan to make long-term earthquake prediction information known to the public at large in relation to the statistical and strain accumulation stages as described in Subsections 5.4.1 and 5.4.2. Designation of an area of intensified observation by CCEP may well correspond to such long-term prediction information.

It may in general be said that the Japanese Government, local governments, public corporations, private enterprises and even individuals behave responsibly regarding such long-term earthquake prediction information. The Japanese Government established the Headquarters for Earthquake Prediction Promotion in 1976 and the Prediction Council for the Tokai area in 1977 in order to cope with not only long-term prediction but also imminent prediction for the expected Tokai earthquake. It should also be noted that the Large-scale Earthquake Countermeasures Act was enacted in December, 1978. The Local Government of Shizuoka Prefecture set up the Anti-earthquake Section in the Prefectural Office and has been pushing earthquake countermeasures of various kinds in close cooperation with the central government.

It is the author's understanding that the area of the southern California uplift is practically under a long-term prediction although it is not quite clear whether the uplift is connected to occurrence of a large earthquake. However, it appears to the author that no tension such as that of local inhabitants in the Tokai area, Japan is felt by the people living in Southern California. This is probably because the seismicity there is substantially lower than that in Japan.

In China, long-term earthquake prediction information is communicated only to executive staff members of revolutionary committees or people's governments of various levels. No such information is made public to local people. This seems also the case in the U.S.S.R. Such a policy would be extremely difficult in Japan, however, because of the highly-developed news-media, so organised that no important information can be concealed.

ROBSON (1978) pointed out that the media such as newspapers, popular magazines, television and radio should feature extensively material on earthquake prediction, earthquake countermeasures and knowledge of earthquakes in general. Such public relations material is already aired and published on an extensive scale in Japan.

According to the SCIENCE AND TECHNOLOGY AGENCY (1978), the reactions of local inhabitants on the release of the possibility of the Tokai earthquake can be summarized as shown in Table 11.1. The Institute for Future Technology selected at random 500 persons as samples for enquiry respectively from the pollbooks at Shizuoka and Shimizu Cities and 250 respectively from those at Fukuroi City and Matsuzaki Town. All these cities and towns are under menace of the feared Tokai earthquake. The percentages of those who replied to the enquiries amounted to 75.8, 75.4, 86.0, and 82.4% for the respective cities and town. The investigation by NHK was also of the same scale.

TABLE 11.1. Appraisals in percentage for the release of the Tokai earthquake hypothesis (SCIENCE AND TECHNOLOGY AGENCY, 1978).

	Good	No good	No comment	Unknown
Investigation by NHK*	43.1	37.6	14.4	5.0
Investigation by Institute for Future Technology	54.1	22.0	22.0	2.0

* NHK is the abbreviation of Nihon Hoso Kyokai (Japan Broadcasting Association).

It is noticeable that 40–50% of local citizens approve issuance of long-term earthquake prediction information.

11.2   Medium- and Short-Term Earthquake Prediction Information

As stated in Subsection 5.4.3., medium- and short-term earthquake prediction information may be put forward corresponding to the stage for the appearance of precursors of the first kind. In recent years it has been customary in Japan that the CCEP President points out detection of an anomalous upheaval at a certain area, if any, at a press conference after a regular or extraordinary meeting of CCEP. Such an announcement seems likely to be understood as a kind of medium- or short-term earthquake prediction information by the mass media although nothing is officially said about earthquake occurrence. The news-media would say that an earthquake of so and so magnitude might occur in so and so area within about so and so years, should the uplift be connected to earthquake occurrence.

It is difficult, however, to judge whether or not the anomaly is really connected to earthquake occurrence in early stage of development of the anomaly. The finding of anomaly may be worth being put on front-pages of newspaper. The news should be continually reported thereafter, depending upon the results of further observations, so that public interest can be maintained.

The report should not be exaggerated. As an earthquake of magnitude 6 or 7 usually occurs several years after the detection of anomaly, mass media should constantly endeavor to attract attention of the public to the earthquake problem.

At this stage, central and local governments should work diligently advising the potential danger of a destructive earthquake to the local inhabitants in the area where the anomaly is observed, together with administrative organs of lower levels, public cooperations and private enterprises in such a way that these organizations can put forward earthquake countermeasures well beforehand. When the local people live off tourism, naturally they are reluctant to accept earthquake prediction information, because they are afraid of loosing guests and sightseers. Even in an ordinary industrial area, some local inhabitants would tend to ignore the information because they do not want to suffer possible economic depression. In spite of these situations, government must direct the local people in such a way that they can cope successfully with the expected earthquake.

11.3   Imminent Earthquake Prediction Information

When precursors of the third or second kinds as defined in Section 4.11 are observed, imminent earthquake prediction information must be made public. According to the experiences in China, this is the stage for which the number of reports on macroscopic anomalies increases enormously. The typical lead time would be several hours to a few days.

Should anomalies exceeding the prescribed level (see Subsection 6.1.5) be observed, and should the Chairman decide to summon the Prediction Council for the feared Tokai earthquake in Japan, the mass news-media will want to immediately report the emergency call of the Council. At the moment, it has been tentatively agreed between JMA and the news-media that news of the emergency call would be released 30 min after the Chairman's decision. It is anticipated, however, that precursory effects of various disciplines would be continually observed even after the stage for short-term earthquake prediction. As appearance of such effects would be followed closely by the mass media, since it seems likely that they are on the alert once the Prediction Council is called.

It seems indeed possible that social confusion of some sort might be caused just by the news that the Prediction Council has been called, although no official conclusion of the Council had yet been put forward. It would also seem likely that information about imminent prediction may leak to the public through personnel working on prediction observation and disaster prevention, via their families or friends. In that case the information is apt to be deformed and so gives rise to over-reaction by local people.

It is the author's belief that such activities regarding the role of the mass media and leakage of information cannot be controlled in a country like Japan where citizens request the right to information, and that the wisest way to deal with imminent earthquake prediction information might be to make everything open to the public,

TABLE 11.2.   Actions to be taken by various organs in Japan in case of issuance of imminent earthquake prediction or earthquake warnings announcement by the Prime Minister (SCIENCE AND TECHNOLOGY AGENCY, 1978).

Administrative organizations belonging to or controlled by central government		Public corporations		Local governments	
Police Agency, Prefectural Police Center and Police Station	Communicate to the locale the earthquake prediction information and related matters by patrol cars, PR cars and policemen.	NHK*	Broadcast the call of Prediction Council. Broadcast the result of Prediction Council meeting as quickly as possible. Broadcast the measures to be taken by the central and local governments and local inhabitants on the basis of the judgment by the Council.	Prefecture Office and its Section of Fire Defence and Disaster Prevention	Order the cities, towns and villages to communicate to the locale the earthquake prediction information and related matters.
Fire Defense Board and Fire Station	Communicate to the locale the earthquake prediction information and related matters by fire engines, ambulance cars, PR cars and firemen. Ask local fire-brigade to join the above communication activity.	Telegraph and Telephone Corporation	Give priority to the telephone circuits for emergency communication.	Prefectural Office	Communicate the information to the staff members by office broadcasting.
National Land Agency, Police Agency and Fire Defense Board (National Headquarters for Earthquake Disaster Prevention)	Communicate to the locale the earthquake prediction information and matters related to prevention of earthquake disaster.			City, town and village offices	Communicate to the locale the information by sirens, fire bells, PR cars, wired and wireless radioes and so on. Contact police stations, volunteer fire-brigades and the like about mutual cooperation.

* Nihon Hoso Kyokai (Japan Broadcasting Association).

although administrators responsible for disaster prevention may not be in favor of such an idea.

It is anticipated in case of the issuance of imminent earthquake prediction information or earthquake warnings statement, that central and local agencies and public corporations in Japan would react as summarized in Table 11.2. In practice, there would be many difficulties: (1) Communications by PR cars is not dependable, (2) There are a number of people who cannot be reached by television and radio, (3) The contents of information can easily be modified, etc.

A large-scale maneuver of earthquake disaster prevention was carried out on November 16, 1979 on the assumption that geophysical anomalies that exceed the prescribed level were observed in the Tokai area. It happened that the Minister of the National Land Agency, who issued the hypothetical earthquake warnings announcement on behalf of the Prime Minister, mentioned that the expected epicenter would be in Suruga Bay. Meanwhile, the Chairman of the Prediction Council said at the press conference, where he was in a position to explain the technical detail of imminent prediction, that the expected epicenter would be somewhere to the south of Omaezaki Point. The author does not know how such a discrepancy occurred. The hypothetical information said that a great earthquake might occur within a few days' time. The information, which was communicated to the local people living in Yaizu City, Shizuoka Prefecture, was that an earthquake would occur within a few hours' time.

It is rather surprising that the information is modified and deformed so easily that even the Minister, who has the highest responsibility, announces a wrong statement. Moreover, this was only a maneuver for training for which everything should have been well prepared beforehand. In a real emergency case, we can easily expect to have much more confusion of communications.

One of the difficulties of communicating imminent information is the point that there are people who do not understand the common language. This is especially the case for the minorities in the U.S.A. In the case of Shizuoka Prefecture, there are usually many tourists staying in hotels and inns in the resort areas. These people are not acquainted with the urgent risk of the great earthquake. Should the great Tokai earthquake occur, tsunami waves would hit the coast of Suruga Bay within 5 to 10 min. In such areas, imminent prediction information is almost equivalent to a warning for evacuation.

At this point, the author should like to refer to the investigations by the Institute of Future Technology (SCIENCE AND TECHNOLOGY AGENCY, 1978) on knowledge and opinion of local inhabitants about imminent earthquake prediction. The investigations were carried out at Shizuoka, Shimizu, and Fukuroi Cities, and Matsuzaki Town in Shizuoka Prefecture as already stated in Section 11.1.

According to the results of investigation, the percentage of people who believe that large earthquake can be predicted several hours before the earthquake occurrence amounts to about 60%. This is the sum of the percentages of people belonging to categories A and B as can be seen in Fig. 11.1.

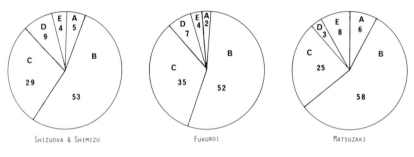

SHIZUOKA & SHIMIZU   FUKUROI   MATSUZAKI

A: POSSIBLE FAIRLY ACCURATELY
B: POSSIBLE TO SOME EXTENT
C: ALMOST IMPOSSIBLE
D: ABSOLUTELY IMPOSSIBLE
E: NO COMMENTS

FIG. 11.1.   Possibility of imminent earthquake prediction in units of percentages (SCIENCE AND TECHNOLOGY AGENCY, 1978).

In Fig. 11.2, we see that the percentage of local people, who wish to have imminent prediction despite the possibility of false predictions, amounts to about 70% or so.

Concerning false prediction information, people are so tolerant that about half of them think that it cannot be helped to have false cases a few times although some 30% people think that a prediction

268

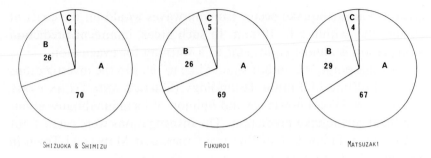

SHIZUOKA & SHIMIZU   FUKUROI   MATSUZAKI

A: IMMINENT PREDICTION INFORMATION SHOULD BE ISSUED POSITIVELY.
B: SHOULD BE ISSUED ONLY WHEN THE PREDICTION IS CERTAIN.
C: OTHER ANSWERS.

FIG. 11.2.   Pros and cons about issuance of imminent earthquake prediction in units of percentages (SCIENCE AND TECHNOLOGY AGENCY, 1978).

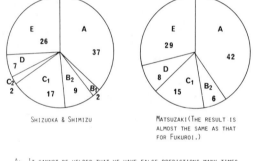

SHIZUOKA & SHIMIZU   MATSUZAKI(THE RESULT IS
                          ALMOST THE SAME AS THAT
                          FOR FUKUROI.)

A: IT CANNOT BE HELPED THAT WE HAVE FALSE PREDICTIONS MANY TIMES.
B_1:   "   MORE THAN 10 TIMES.
B_2:   "   4 TO 5 TIMES.
C_1:   "   2 TO 3 TIMES.
C_2: IT CANNOT BE HELPED THAT WE HAVE ONE FALSE PREDICTION.
D: NO COMMENTS.
E: EARTHQUAKE PREDICTION SHOULD BE ISSUED ONLY WHEN IT IS CERTAIN.

FIG. 11.3.   Allowance for swing-and-miss prediction information in units of percentages (SCIENCE AND TECHNOLOGY AGENCY, 1978).

should be issued only when it is certain, as can be seen in Fig. 11.3. About 50% of local inhabitants know the existence of a system for prediction although very few of them know the exact name of the system. This situation is illustrated in Fig. 11.4.

In China, imminent earthquake prediction information or an

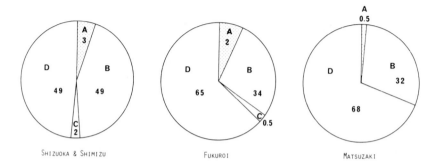

A: Knowing the name of Prediction Council.
B: Knowing the existence of such a system.
C: No answers.
D: Knowing nothing about such a system.

Fig. 11.4. How well is the Prediction Council known? The numerals indicate percentages (Science and Technology Agency, 1978).

earthquake warning is issued from a Headquarters for Earthquake Disaster Prevention established at the locale by order of the Revolutionary Committee as shown in Table 8.1. Should the warning time permit, the information may be made public by the so-called "wall-newspapers" at people's communes. In practice, telephone communication seems to be important. In the case of the Longling earthquakes ($M = 7.5$, 7.6, 1976), the warning was conveyed to the public by sirens only 25 minutes prior to the earthquake occurrence. As television and radio are not quite as popular in China to the extent of western countries, the authorities do not put much stress on communication by these media.

11.4   Communication Problem of Earthquake Information in the Case of the Izu-Oshima Kinkai Earthquake

No imminent earthquake prediction information has ever been issued except for a few examples in China, so that it is not known what would happen in an actual case in a country of liberalism such as Japan or the U.S.A. As quoted in Table 8.2, however, urgent earthquake information was released by JMA in relation to the swarm activity near the Izu-Oshima Island which was immediately followed by the Izu-Oshima Kinkai earthquake ($M = 7.0$, 1978). Furthermore, the Governor of Shizuoka Prefecture made the public aware of the "aftershock information" which unexpectedly gave rise to social panic or confusion of some sort. Although these examples are not necessarily called predictions, they provide good instances of tumult caused by misunderstanding information while being transcribed or communicated. It is therefore planned in the following to describe the main points of difficulty in the course of transmitting the earthquake information related to the Izu-Oshima Kinkai earthquake. The following are largely due to the SCIENCE AND TECHNOLOGY AGENCY (1978), SHIMAZU and HIRAMATSU (1978) and the INSTITUTE OF JOURNALISM (1978).

11.4.1 Transmission of the swarm earthquake information
    The progress of seismicity before and after the JMA earthquake information as shown in Table 8.2 has been briefly mentioned in Section 8.5. The scientific and social events related to the Izu-Oshima Kinkai earthquake are summarized in Table 11.3.
    The issuance of earthquake information was sent by JMA to the Department of Disaster Prevention of Tokyo Prefectural Office and other organs amounting to about 80 in number with a special instrument for simultaneous transmission. On the other hand, a telex was sent to the Shizuoka Meteorological Station at the same time. A press conference was held at JMA at 11 : 00 when the background of information was explained in detail.
    The earthquake information thus issued was transmitted to the public through the channels of administration and mass media as shown in Fig. 11.5. The Shizuoka Meteorological Station reported on the information to the Section of Fire Defense and Disaster Prevention, Shizuoka Prefectural Office at around 11 : 00. The Section of Fire Defense and Disaster Prevention sent out the information to all cities, towns and villages in the prefecture just as they do whenever they receive earthquake information. As the earthquake activity was thought to be occurring in a limited area around the Izu-Oshima Island, they did not feel any tension and so no specific counter-measures were instructed. The transmission reached cities, towns and

TABLE 11.3. Scientific and social events related to the Izu-Oshima Kinkai earthquake (SCIENCE AND TECHNOLOGY AGENCY, 1978).

Date	Time	Event
Jan.13	Evening	Swarm earthquakes began to occur in the sea area near the Izu-Oshima Island.
14	9:00–10:00	Many earthquakes were felt in the Izu-Oshima Island.
	10:50	JMA issued "earthquake information" that mentioned the possibility of a slightly damaging shock. The information was conveyed to the Prefectural Offices of Tokyo and Shizuoka.
	11:15	Shizuoka Prefectural Office conveyed the above information to all the cities, towns and villages in the prefecture.
	Around 12:00	NHK and other commercial broadcasting stations broadcast the earthquake information.
	12:24	The Izu-Oshima Kinkai earthquake of magnitude 7.0 occurred. Many aftershocks followed.
	14:10	Head Office of Shizuoka Prefecture for the earthquake disaster was set up. Similar head offices were set up at 2 cities and 5 towns in Shizuoka Prefecture during 12:45–14:50.
15	7:31	An aftershock having the maximum magnitude of 5.8 occurred at the central part of the Izu Peninsula.
	18:00	CCEP publicized its official view that aftershocks would continually occur over a period of half a month and that the maximum shock might have a magnitude of 6 or thereabout. The view was circulated to the public by television, radio and newspapers.
16	10:00	Head Office for the Disaster Measures of the Izu-Oshima Kinkai earthquake was set up in the Japanese Government.
17	Evening	The above Head Office of the central government announced at a press conference the prospect of earthquake activity following the CCEP's view.
18	9:30	The Governor of Shizuoka Prefecture asked his staff members to form the "aftershock information" to be distributed in the prefecture.
	13:30	The draft of "aftershock information" was completed by the prefectural head office.
	13:40	The Governor announced at a press conference the "aftershock information". At the same time, the information was sent to cities, towns and villages in the prefecture by means of the wireless communication system.

272

TABLE 11.3 (continued)

Date	Time	Event
Jan. 18	14:17	SBS (Shizuoka Broadcasting Station), a commercial broadcasting station, radioed the "aftershock information". The same news was repeated at 14:40 and 15:00.
	14:34	SBS announced twice the news on TV with a simplified sentence superposing on the images of regular program.
	14:55	Similar broadcasting by the TV Shizuoka.
	15:05	NHK Shizuoka Station radioed the "aftershock information" in detail.
	14:00–15:30	At about 20 cities, towns and villages mostly in the Izu Peninsula, the "aftershock information" was conveyed to the public by wired radios and PR cars.
	15:55	SBS TV broadcast the "aftershock information" as a part of regular news program.
	14:00–17:30	There was a rush of enquiries about the "aftershock information" to the prefectural, city, town and village offices, fire departments, police stations and offices of mass media.
	16:30	The Governor advised that there was no immediate danger in order to quiet the confusion caused by misunderstanding of information. What he said was conveyed to all the cities, towns and villages in the prefecture.
	16:55	TV Shizuoka broadcast in a special news program new information which denies the rumor caused by misunderstanding.
	Around 17:30	NHK TV and radio broadcast similar messages several times during the Sumo (Japanese wrestling) program.
	17:30	Head Office of Shizuoka Prefecture for the earthquake disaster asked all cities, towns and villages to watch and listen to SBS TV at 18:00 and NHK TV at 18:40. The confusion began to ebb.
	18:00	The General Manager of the Prefectural Office appeared on SBS TV and explained the real intention of the "aftershock information".
	18:40	The General Manager did the same thing on NHK TV. All TV and radio stations reported the events related to the "aftershock information". All the confusion seemed to die down by this time.
19		Morning newspapers reported the panic and confusion in detail.
Feb. 7		CCEP announced that the aftershock activity had almost diminished.

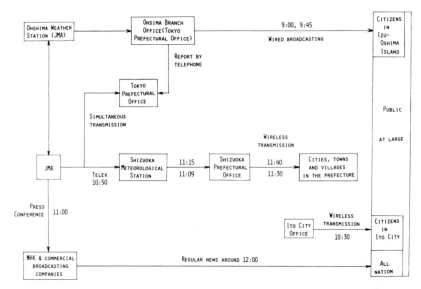

Fig. 11.5. The flow of earthquake information of the swarm seismic activity around the Izu-Oshima Island on Jan. 14, 1978 (SCIENCE AND TECHNOLOGY AGENCY, 1978).

villages at around 11 : 40. Being a Saturday, the working hours finished at noon. As nobody thought that an earthquake, that would induce some damage on the Izu Peninsula, would occur, no public relations activity and countermeasures for earthquake disaster was undertaken.

The only exception was the Ito City Office which issued an earthquake warning of some sort at around 10 : 30 on the basis of independent judgment. The judgment was made based on the seismogram of a simple seismograph installed at the Fire Station of the city.

The Oshima Branch Office of the Tokyo Prefectural Office had been contacting very closely with the central office. The Branch Office, which is situated very close to the epicentral area of earthquake swarm activity, had already asked the local people not to use petroleum stoves and to shut down valves for fuel gas at 09.45. The Tokyo Prefectural Office received the information from JMA, but they took no specific action.

NHK and other commercial broadcasting companies broadcast

the earthquake information by TV and radio as one of the items of nation-wide and local news. As a result, substantially more people received the news via TV-radio channels rather than administrative public relations activity, although very few people working in offices, shops or factories, could be reached by the TV-radio information. And this information did not produce much public reaction, no matter in which form it was received, because no sense of immediate concern or tension was conveyed from the news. However, this was of course not the case in the Izu-Oshima Island. Some people living in the Izu Peninsula thought that the activity was limited to the Izu-Oshima area only and so they tended to be off their guard.

Judging from the reaction of local inhabitants as disclosed by the present earthquake information, it is sometimes useless to convey information unless some specific directions about appropriate actions to be taken by local people are involved. It is highly necessary that leaders of neighborhood association, workers' groups and so on display their leadership.

11.4.2 Aftershock information and related problems

(1) Issuance of the aftershock information

On the day following the occurrence of the Izu-Oshima Kinkai earthquake, that is January 15, CCEP had a meeting of its Kanto Section. After investigating all the information relevant to the present seismic activity, CCEP released its official view that is indicated in Table 11.4.

At the January 17 meeting of the Head Office for the Disaster Measures of the Izu-Oshima Kinkai earthquake, a prospect about the current seismic activity as shown in Table 11.5 was adopted on the basis of the CCEP's view in Table 11.4. Unlike the CCEP's view, that written in Table 11.5 ignored the probable duration of aftershock activity and emphasized anticipated damage by earthquakes.

The Governor of Shizuoka Prefecture is not in a position to be officially notified by either CCEP or Head Office for the Disaster Measures of the Izu-Oshima Kinkai earthquake because of the present administrative organization of the Japanese Government. As a consequence, he could be informed only unofficially about the above views. The Governor, who took notice of the fact that

TABLE 11.4.   The official view of the Kanto Section of CCEP about the Izu-Oshima Kinkai earthquake on Jan. 15, 1978.

The earthquake that occurred at 12:24 on Jan. 14 was located at a point about 10 km west of Izu-Oshima Island.

According to the observation by JMA, the earthquake was of right-lateral type and the associated fault struck approximately in an east-west direction. The earthquake magnitude was estimated as about 7. This was a shallow earthquake having the largest magnitude for the area.

Aftershocks are occurring not only beneath the sea around Izu-Oshima Island but also in the middle part of Izu Peninsula. It is not quite clear whether these shocks occurring in the peninsula are ordinary aftershocks or an independent seismic activity which was induced by the Jan. 14 earthquake.

At any rate, it is expected that the seismic activity would last for some time, so that the local inhabitants living in Izu Peninsula are requested to be on the alert probably for half a month. The maximum magnitude expected for the largest aftershock would be 6 or thereabout.

It is difficult to say whether the present seismic activity is directly connected to the so-called "Tokai earthquake" and other great earthquakes that occur in the Sagami Bay. No indication that the shock is connected to an eruption of Volcano Mihara on Izu-Oshima Island had been found.

TABLE 11.5.   Probable development of the seismic activity in the immediate future as disclosed by the Head Office for the Disaster Measures of the Izu-Oshima Kinkai earthquake on Jan. 17, 1978 based on the official view of the Kanto Section of CCEP.

It is said that energy amounting to one-tenth of that of the main shock at the most may be released by an aftershock. In the case of the Izu-Oshima Kinkai earthquake, only 1/40 of the energy of the main shock has been released by the aftershocks as of now. In the worst case, therefore, it is possible to have an aftershock of magnitude 6 or thereabout (equivalent to the Ebino earthquake in 1968). As the focus is expected to be shallow, moderate to severe damage for an $M = 6$ earthquake might be caused.

aftershocks had been occurring in his prefecture and that the main shock had given rise to considerable damage, made up his mind to issue "aftershock information" on his own responsibility. On the basis of the draft formed by the Prefectural Office under the Governor's direction, the Governor issued the "aftershock information" to the public at large in his prefecture and, at the same time, held a press conference at 13:40 on January 18. The full text of the aftershock information is reproduced in Table 11.6.

TABLE 11.6.   Aftershock information issued by the Shizuoka Prefecture on Jan. 18, 1978.

Notification of aftershock information

By the Governor

1. The following view was released by the Head Office for the Disaster Measures of the Izu-Oshima Kinkai earthquake on Jan. 17.

"In the worst case, an $M6$ earthquake may possibly occur during the aftershock activity of the Izu-Oshima Kinkai earthquake. As the earthquake foci in the area are in general shallow, moderate to severe damage may well be expected for an $M6$ earthquake in that case."

2. As the restoration work related to the main shock is going on, and we had rain and snow fall on Jan. 18, we should be continuously on the alert in preparation for a new shock.

3. The Prefectural Office is going to take the following measures. The cities, towns and villages in the southern and middle parts of the Izu area should also take necessary action taking the local condition into account.

The Governor hopes that local citizens would behave calmly paying attention to coming information.

    (1) Collection and communication of information

    (2) Intensification of watching system for temporarily repaired roads, railways and the like

    (3) Intensification of patrol over dangerous areas

    (4) Traffic control in dangerous areas

    (5) Preparation of necessary commodities

    (6) Appeal to local citizens

        i)   Caution for fire and emergency measures for strengthening houses

        ii) Voluntary control of traffic

        iii) Preparation for evacuation and confirmation of the roads and places for evacuation

        iv) Preparation for drinking water and emergency food

(2)   Propagation of the aftershock information

(2a)   Channels through public relations activity of cities, towns and villages

The Shizuoka Prefectural Office transmitted the "aftershock information" to the 75 cities, towns and villages in the prefecture and the prefectural branch offices for civil engineering, finance, harbor control and the like by the wireless system of disaster prevention immediately after the Governor's press conference. Telephone communication was also used in some cases.

The 25 cities, towns and villages in the eastern Shizuoka Prefecture, especially those in Izu area, communicated the information to the public through wired broadcasting, simultaneous wireless broadcasting, from public relations cars and fire engines and so on at 14:00–15:00 hours. But the contents of public relations

TABLE 11.7. Aftershock information from the Matsuzaki Town Office (SCIENCE AND TECHNOLOGY AGENCY, 1978).

This is to inform the local inhabitants of aftershock activity from the Town Office.

The following view was put forward at yesterday's meeting of the Head Office for Disaster Measures of the Izu-Oshima Kinkai earthquake. "It is possible to have an aftershock of magnitude 6, in the worst case, in relation to the Izu-Oshima Kinkai earthquake. As the focus is expected to be shallow, moderate to severe damage for an $M=6$ earthquake might be caused." As the restoration work related to the shocks occurred on Jan. 14 and 15 is proceeding, and we expect rain fall hereafter, we should be successively on the alert for a new shock. It is hoped that all the inhabitants pay strict attention to the coming information. In the event an earthquake does occur, please behave calmly.

For the moment, the local people should pay attention to fire prevention. The Town Master once more asks the citizens to inspect the situation of houses and furniture. It is important to mend houses and fix furnitures in such a way that they do not fall in case of a strong shock.

It is urgent for local people to confirm the places of evacuation and the roads that lead to those places.

It is absolutely prohibited to leave by car in the event of a strong shock.

Each family should prepare some water for drink and food for emergency use.

TABLE 11.8. Aftershock information from the Higashi-Izu Town Office (SCIENCE AND TECHNOLOGY AGENCY, 1978).

This is to inform the local inhabitants of aftershock activity.

As the aftershocks of the Jan. 14 earthquake are still occurring and the weather is not good, local citizens are requested to watch dangerous spots around your houses. At the same time, you are requested to be on the alert just in case of another disaster by paying attention to necessary evacuation arrangements and also to securing necessary commodities.

The toll roads are under repair and reconstruction. As there may well be the danger of landslide, you are asked not to take to the roads by car.

statements were different from government to government. Tables 11.7 and 11.8 indicate the information publicized to local inhabitants at Matsuzaki and Higashi-Izu Towns, respectively. The fact that the information is different, one area from another, may easily cause some confusion when people living in a certain district talk over the telephone with others who live in a different district. Suspicion begets fear.

(2b)   Channels through TV and radio

According to the SCIENCE AND TECHNOLOGY AGENCY (1978), the Governor held the press conference in a strained atmosphere. The mass media appreciated the issuance of "aftershock information" because this was the first time in history for a local government takes such a decision, although the contents of information are almost the same as those of CCEP or Head Office for the Disaster Measures of the Izu-Oshima Kinkai earthquake (see Tables 11.4 and 11.5). Judging from the degree of tension and preparation of Prefectural Office, some pressmen conjectured that the Governor might know something more about firm evidence for a coming strong shock although it could not be officially announced.

The first reports about the "aftershock information" were broadcasted by SBS radio at 14:17, 14:40, and 15:00. The contents

TABLE 11.9.   Radio news from the SBS (Shizuoka Broadcasting Station) at 14:17, 14:40 and 15:00 (SCIENCE AND TECHNOLOGY AGENCY, 1978).

The Shizuoka Prefectural Office issued "aftershock information" at 13:30. According to the information, it is warned that, in the worst case, a shock of magnitude 6 might possibly occur in relation to the aftershock activity of the Izu-Oshima Kinkai earthquake. The cities, towns and villages in the southern and middle areas of Izu district are requested to take necessary action. Those who are living within these areas are also requested to pay strict attention to subsequent information that follows and to behave calmly in case of a large earthquake.

The following opinion was released yesterday by the Head Office for the Disaster Measures of the Izu-Oshima Kinkai earthquake. "In the worst case, an $M6$ earthquake may possibly occur during the aftershock activity of the Izu-Oshima Kinkai earthquake. As the earthquake foci in the area are in general shallow, moderate to severe damage may well be expected for an $M6$ earthquake in that case." As we expect rain and snow falls that might lead to new disasters in case of a strong shock, we should be continuously on the alert in preparation for a new shock.

Accordingly, the Prefectural Office is going to take the following measures; (1) Collection and communication of information, (2) Intensification of watching system for temporarily repaired roads, railways and the like, (3) Intensification of patrol over dangerous areas and so on. The office also asks the cities, towns and villages in the southern and middle parts of Izu district to take necessary action taking the local conditions into account.

Local citizens are requested to pay special attention to the following points; (1) Caution for fire, (2) Voluntary control of traffic, (3) Preparation for evacuation and confirmation of the roads and places for evacuation, and (4) Preparation for drinking water and emergency food.

TABLE 11.10. TV news from the SBS superimposed on an ordinary program at 14:34 (SCIENCE AND TECHNOLOGY AGENCY, 1978).

The Shizuoka Prefectural Office issued the "aftershock information" to the southern and middle areas of Izu district this afternoon. Local citizens are requested to pay attention to information that follows and to behave calmly in case of a strong shock.*.

* The broadcasting was repeated twice.

TABLE 11.11. Regular TV news from TV Shizuoka at 14:55 (SCIENCE AND TECHNOLOGY AGENCY, 1978).

Mr. Yamamoto, the Governor of Shizuoka Prefecture, issued the "aftershock information" to the southern and middle areas of Izu district at 13:30 today for the first time in history.

This is because of information put forward at yesterdays' meeting of the Head Office for the Disaster Measures of the Izu-Oshima Kinkai earthquake. The information reads that, in the worst case, an M 6 earthquake might well occur in relation to the aftershock activity.

This is really the first time for a local government to release such aftershock information which is practically equivalent to earthquake prediction information. All cities, towns and villages concerned in the prefecture are working urgently advising local citizens of the latest information.

of the news reproduced in Table 11.9. SBS TV superimposed the simplified sentence of the news as shown in Table 11.10 on its ordinary program at 14:34. The news was repeated twice at that time, and the same news broadcasted at the time of the 15:00 regular news.

On the other hand, TV Shizuoka (SUT) put the news (see Table 11.11) on the air at 14:55 as a part of regular news broadcasting. it was stressed that this was the first time for a local government to issue such information.

The NHK radio sent out the news quoted in Table 11.12 as a part of the local regular news broadcast at 15:05. Neither special radio news nor TV news were put on the air by NHK.

Two wired TV stations in Shimoda City broadcasted the after-shock information at the request of the City Office.

(2c) Channels through propane gas suppliers

Propane gas is widely used for fuel in Shizuoka Prefecture. As the gas is highly inflammable, the Section of Fire Defense and Disaster Prevention of the Prefectural Office transmitted the after-

TABLE 11.12.   Regular radio news about aftershock information NHK Shizuoka Station (SCIENCE AND TECHNOLOGY AGENCY, 1978).

The Shizuoka Prefectural Office issued the "aftershock information" to the southern and middle areas of Izu district. The local citizens living in the dangerous areas are requested to be prepared for evacuation, to secure drinking water and emergency food and so on.

On the basis of the official conclusion of the Kanto Section of CCEP, yesterday's meeting of the Head Office for the Disaster Measures of the Izu-Oshima Kinkai earthquake held in Tokyo adopted the following opinion:

"In the worst case, an $M6$ shock may possibly occur during the aftershock activity of the Izu-Oshima Kinkai earthquake. As the earthquake foci in the area are in general shallow, moderate to severe damage may well be expected for an $M6$ earthquake in that case".

As the Shizuoka Prefectural Office, which has been in close contact with the Head Office for the Disaster Measures of the Izu-Oshima Kinkai earthquake, judged that the area should further be on the alert, the present "aftershock information" is issued.

The local citizens living in the southern and middle areas of Izu district are therefore requested to pay strict attention to the following points:

>Caution for fire
>Voluntary control of traffic
>Preparation for evacuation and confirmation of the roads and places for evacuation
>Preparation for drinking water and emergency food
>Emergency measure for strengthening houses

Mr. Yamamoto, the Governor, said "Although the information might frighten local citizens to some extent, I really feel that it is necessary for the southern and middle areas of Izu district to be on the alert"

---

shock information to the Association of Propane Gas Suppliers by telephone. The information was communicated to the regional and local gas stations through the special channels of the association. The information was picked up by local people by chance during the process of propagation, and as will later be mentioned, it was deformed to a great extent resulting in much confusion.

(2d)   Channels through the Prefectural Board of Education Committee and other Sections

The Board of Education of Shizuoka Prefecture was ordered to communicate the aftershock information to high, middle and primary schools in the prefecture. It turned out, however, that it took some 3 to 4 hours in order to complete passing such information because of the telephone communication system for which it was planned to send information successively from a school to school. The aftershock information sent out from various Sections of the Prefectural Office reached health offices, prefectural hospitals, associations of medical doctors, associations of commerce and industry, prefectural tourist bureaus, bus companies, road, telephone and telegram corporations and the like at around 14:00 to 15:00 hours.

(3)   Spreading and modification of the aftershock information

It became clear that the TV-radio channel was the most efficient for the purpose of communicating the information. Just over one third of the total population of Shizuoka Prefecture received firsthand information; the 80% of that total were informed by TV and/or radio.

It appears that a large proportion misunderstood the information even though they were reached by it. According to the SCIENCE AND TECHNOLOGY AGENCY (1978), the ratio of misunderstanding for various media are those as given in Table 11.13. Ratio of misunderstanding X is defined by

$$X = \frac{A}{B-C} \times 100 \, (\%)$$

in which A is the number of people who misunderstood the information, B the number of people who received the information, and C the number of people who did not pay special attention to the information received.

Judging from the above result, it is highly important for mass media to improve contents, expressions and timing of news broadcasting in case of emergency in such a way that the information may be correctly communicated to local people. Especially, attention

TABLE 11.13.   Rate of misunderstanding for various media as estimated for the prefecture as a whole (SCIENCE AND TECHNOLOGY AGENCY, 1978).

NHK radio news	37.5%
Extraordinary radio news of commercial broadcasting	43.8
TV news of commercial broadcasting	48.3
Wired broadcasting in cities, towns and villages	51.9
PR cars · fire engines	52.0

should be particularly drawn to the point that no exaggerated anxiety should be caused.

Typical misunderstandings are as follows;

i) A great earthquake occurs shortly.

ii) An earthquake warning is issued.

iii) An earthquake of intensity VI would soon occur. This is probably due to the misunderstanding of the word "$M\,6$" involved in the information.

iv) A large earthquake would occur in the southern or middle parts of the *prefecture* instead of the *Izu district*.

The information was further modified with additional demagogy. Typical modifications are as follows;

i) A great tsunami would hit.

ii) An earthquake would occur some time between 4 and 6 o'clock in the afternoon.

iii) Volcano Mihara on Izu-Oshima Island is about to erupt.

iv) A pressman ran away aboard a helicopter. Many people have left their living places for evacuation places.

Much of such false speculation and rumor was formed during transmission of official information propagated secondhand and distorted from person to person. Some 57% of the people in the prefecture learned of such rumor or misinformation from second-hand-sources. Such a percentage amounted to 1,900,000 people.

The modification of information took place for all channels as discussed in (2a), (2b), (2c), and (2d) in this subsection. It is clear, however, that the information was dreadfully distorted in the early stage of the propagation through the propane gas suppliers channel.

It is not only too tedious for this book to discuss the process of information modification and its cause but also such an undertaking that it is a matter beyond the author's ability. Those who are particularly interested in this aspect of the aftershock information of the Izu-Oshima Kinkai earthquake are kindly requested to refer to SCIENCE AND TECHNOLOGY AGENCY (1978), SHIMAZU and HIRAMATSU (1978) and the INSTITUTE OF JOURNALISM (1978) although the author points out that all these papers are written in Japanese.

It becomes clear all through the investigations of communication about the aftershock of the Izu-Oshima Kinkai earthquake that the information required, as a first priority by local citizens, is the "time of earthquake occurrence". When an earthquake warning is to be issued, information about the occurrence time should be clearly indicated.

11.5   Models of Public Broadcasting in Relation to Imminent Earthquake Prediction Information

It has been demonstrated in Sections 11.3 and 11.4 that the TV-radio channel is the most efficient and effective for the purpose of conveying imminent information of earthquake prediction to the public at large. As there may be many difficulties, however, in actual broadcasting, NHK, which is the official broadcasting medium for disaster prevention in Japan, has been seeking the best way of public broadcasting in case an imminent prediction of the feared Tokai earthquake should be made.

In the following, the author takes the liberty of quoting the provisional manuscripts prepared by the NHK News Section (M. Kimura, personal communication, 1979) as models of public broadcasting related to an imminent prediction of the Tokai earthquake. It should be borne in mind, therefore, what follows are mere models of hypothetical broadcasting as arranged according to time series, and it is highly likely that they are modified in the course of future development of anti-earthquake measures.

11.5.1   Summons of Prediction Council
This is to broadcast an extraordinary news item.

Anomalies were detected in the data for earthquake prediction in the Tokai area. In order to judge whether the anomalies are connected to occurrence of a great earthquake, it has been decided to call a meeting of experts at JMA.

According to JMA, anomalies were found in the data for earthquake prediction in the Tokai area so and so hours today. The Prediction Council, consisting of 6 experts, has organised a meeting at JMA in order to investigate whether or not the anomalies have a potential towards the occurrence of a great earthquake.

The members of Prediction Council will shortly arrive at JMA. It is expected that they would reach a conclusion as to the occurrence of a great earthquake about 30 minutes after the meeting begins at the latest.

NHK will hereafter broadcast the information about the Prediction Council meeting for the Tokai earthquake by interrupting all other broadcasting. Please pay attention to the NHK broadcasts from now on.

11.5.2 Watch and listen to NHK broadcasts to obtain further information about the Tokai earthquake

This is to repeat the information about the summons of the Prediction Council for the Tokai earthquake.

Anomalies were detected in the data for earthquake prediction in the Tokai area. In order to judge whether the anomalies are connected with the occurrence of a great earthquake, it has been decided to call a meeting of the Prediction Council.

Even though the Council has been called, it is not known whether a great earthquake will occur. When the details of the data observed or the conclusion of the Council are released, NHK TV-radio will immediately broadcast that news.

Please wait for the information that follows.

11.5.3 About radio-TV channels for the emergent broadcasting

This broadcast simultaneously transmitted by NHK radio 1, 2 and FM, NHK TV general and education.

11.5.4 All about the Prediction Council

The Prediction Council for the Tokai earthquake, which is officially called the Prediction Council for Areas under Intensified Measures against Earthquake Disaster, is a group of experts of seismology. The Council is an advisory organ to JMA's Director General.

The Chairman is Professor Toshi Asada, a professor of Tokai University. The members are as follows;

Professor Tsuneji Rikitake, Nihon University,

Professors Kiyoo Mogi, Keichi Kasahara, Tatsuo Usami, and Tokuji Utsu, Earthquake Research Institute, University of Tokyo.

The Council is in a position to judge whether or not a great earthquake occurs on the basis of incoming data telemetered from the observation arrays over the Tokai area to JMA.

The Council meeting begins as soon as all the members arrive at JMA. It is expected that the judgment will be made within 30 minutes' time Council convening at the latest.

11.5.5 When will a meeting of the Prediction Council be held?

The anomalies observed this time will shortly be released from JMA. It has been provisionally decided that the Council may be called when the anomalies exceed one of the following prescribed levels;

(1) A seismograph records 10 or more swarm earthquakes per hour, including 3 or more earthquakes having magnitude 4 or over, towards the west of the Suruga trough and such a state continues 2 hours or longer along with remarkable changes recorded by 2 or more volume strainmeters at about the same time.

(2) A volume strainmeter records a rapid expansion or contraction of crustal rocks exceeding a prescribed limit and conspicuous changes of similar kind are observed by at least 3 more volume strainmeters at about the same time (see 6.1.6).

These anomalies are not necessarily connected to the occurrence of a great earthquake. Even with anomalies which are thought to be an earthquake precursor, there are cases for which no earthquake occurs. On the contrary, an earthquake may occur without being forerun by a conspicuous premonitory effect.

Accordingly, even though the Prediction Council puts forward a judgment that it is highly likely that a great earthquake would occur, it is possible that no earthquake actually occurs.

11.5.6 Probable development from now

In case the Council should judge that the Tokai earthquake would soon occur, the earthquake prediction information would be reported to the Prime Minister via JMA's Director General. The Prime Minister would then issue an "earthquake warnings statement" upon consultation with the cabinet. In that case, various organizations such as hospitals, theaters, department stores, hotels, gas stations, petroleum refineries, private railways and the like in the "area under intensified measures against earthquake disaster" would be prepared for the earthquake in accordance with the "short-term plan of earthquake disaster prevention" which has been previously established. The said area involves all Shizuoka Prefecture and some parts of Kanagawa, Yamanashi, Nagano, Gifu and Aichi Prefectures.

The above is the probable procedure in case the Council should judge that the occurrence of a great earthquake is highly likely.

On the other hand, if it is concluded that the anomalies do not indicate the occurrence of a great earthquake, everything will resume its daily routine.

In any case, whether the present anomalies are connected to earthquake occurrence is not as yet known.

11.5.7 Please wait calmly for further information

In the "area under intensified measures against earthquake disaster" and its surroundings, police and fire-brigades will be prepared for the earthquake just in case a great earthquake would occur. Should a positive conclusion that a great earthquake will occur be agreed by the Prediction Council, city, town, and village offices will issue practical guidelines for evacuation and other necessary measures over areas where the danger of landslides and tsunamis is high.

It is at the moment not known whether or not a great earthquake will hit. No detailed information has as yet been received by police, meteorological stations and the like. This is the stage that a meeting of the Prediction Council has been called at JMA in Tokyo.

The conclusion of the Council meeting will shortly be broadcast by radio and TV. Please wait for further information in a calm manner.

11.5.8 The Prediction Council meeting has started

The meeting of the Prediction Council has just started at JMA.

The judgment whether or not the Tokai earthquake would occur will shortly be released.

Professor Asada, the Chairman of the Council, and other members have arrived at JMA one by one.

A short time ago, the meeting started at so and so minutes and so and so hours. The Council is trying to reach a conclusion whether the Tokai earthquake will occur as quickly as possible by analysing all the data available.

When the conclusion is released, NHK will immediately put it on the air.

11.5.9   Conclusion of the Prediction Council

The Prediction Council meeting seems to have reached a conclusion. It will shortly be announced at a press conference. Please wait a few moments without switching the NHK TV-radio off.

11.5.10   Earthquake prediction information

The "earthquake prediction information" that the Tokai earthquake would occur within so and so hours from now has been issued by JMA basing on the conclusion of the Prediction Council.

The expected earthquake magnitude would amount to so and so which is almost equal to that of the 1923 Kanto earthquake. The time of earthquake occurrence cannot be specified exactly, but the possibility of the earthquake occurring within so and so hours from now, namely between now and so and so hours, is estimated to be high.

Should the earthquake occur, the "area under intensified measures against earthquake disaster", that is the area centering on Shizuoka Prefecture and including some portions of Kanagawa, Yamanashi, Nagano, Gifu, and Aichi Prefectures, would be shaken by extremely strong ground vibrations of intensity VI or stronger, so that it is feared that buildings and houses will be destroyed and that landslides occur.

It is also anticipated that a large tsunami of which the wave height exceeds 3 m would hit the coast of Suruga Bay from Shimoda on Izu Peninsula to Omaezaki Point.

Even outside the above area under intensified measures, strong ground vibrations amounting to intensity V would hit the Tokyo-Yokohama area and the Nobi area including Nagoya.

The earthquake prediction information will soon be reported to the Prime Minister by JMA's Director General. The Prime Minister will shortly issue the "earthquake warnings statement".

11.5.11   Announcement of earthquake warnings statement

This is to broadcast the "earthquake warnings statement" which was just now announced by the Prime Minister.

All administrative organs in the "area under intensified measures against earthquake disaster" are requested to carry out the "short-term plan of earthquake disaster prevention" which has already been established.

Those who are in the "area under intensified measures against earthquake disaster" should stop the use of domestic fire and refrain from driving a car. They should also be prepared for evacuation and fire-fighting. Please behave calmly under the direction of local governments and police.

Those who are outside the "area under intensified measures against earthquake disaster" are asked to cooperate with the central and local governments by not going to the area concerned and obeying the traffic control. Do not call someone in the area by telephone because telephone lines are much too busy.

In his "earthquake warnings statement", the Prime Minister asks the nation to cooperate with him by paying attention to the above points.

11.5.12   Establishment of headquarters for earthquake disaster prevention

The National Headquarters for Earthquake Disaster Prevention was just established in preparation for the Tokai earthquake. The Headquarters is headed by the General Director. The Prime Minister becomes the General Director. The members of headquarters are ministers of various ministries and agencies concerned and representatives of specified public corporations such as Japan National Railways (JNR), Telephone and Telegram Corporation, NHK and so on. The National Headquarters assumes responsibility for enforcement and coordination of the short-term plan of earthquake disaster prevention for the Tokai earthquake.

In the "area under intensified measures against earthquake disaster" prefectural and municipal headquarters are empowered to carry out the short-term plan of earthquake disaster prevention proper to respective prefectures, cities, towns and villages by performing traffic control, evacuation instruction and the like.

11.5.13   Start of short-term plan of earthquake disaster prevention

In conjunction with the establishment of the National Headquarters for Earthquake Disaster Prevention, the short-term plan of earthquake disaster prevention has been started in and around the "area under intensified measures against earthquake disaster".

First of all, the following traffic control is brought to effect:

Vehicle traffic is restricted on main roads in and around the "area under intensified measures against earthquake disaster". That is also the case for roads for evacuation in the area concerned. The Headquarters requests the citizens not to use a car.

The Tomei High Way (connecting Tokyo to Nagoya) is closed.

The Shinkansen (bullet train) between Tokyo and Nagoya is stopped completely. As for the Tokaido line and other lines of JNR in and around the area concerned, these may be stopped or operated with a slow speed.

Electricity is supplied as usual. However, citizens are asked to prepare transistor radios and flashlights by way of precaution.

The pressure of gas for fuel will become lower than that at ordinary time. As flames tend to become smaller and subjected to blow out by wind, much caution is required.

Haneda, Tokyo and Komaki, Nagoya airports will be closed.

The following concerns evacuation. Instruction for evacuation is issued from local headquarters for earthquake disaster prevention in an area where extremely strong shocks are expected. This is also the case for areas which is under the threat of tsunami and landslide. In such a case, please evacuate calmly under the instruction of policemen or firemen.

Safety measures for children are taken at schools in accordance with the prescribed plan.

Short-term plans such as mentioned above will be hereafter put on the air for each district respectively from the NHK branch stations in the locale. Please watch and listen to TV-radio information that follows.

11.5.14   All about citizens' countermeasures against earthquake disaster

The earthquake prediction information that the Tokai earthquake will occur within so and so hours from now with a high probability has been released from JMA. If we are well prepared for the earthquake, we may be able to minimize the earthquake damage. Be prepared for the earthquake at your homes and your places of work.

(1)   Ten key-points at home.

i)   Let's have a family meeting for disaster prevention. Who is going to extinguish fire? Who is going to take care of children and aged persons? Decide the jobs of each family member well beforehand.

ii)   Examine the safety in your home. Where is the safest place in your home? Think over the place where you go for the time being in case of strong shock. Have you fixed heavy furniture to the wall? Heavy articles or glass products which may fall at the time of earthquake must be taken down immediately. If possible, make your house strong by making use of diagonal beams and the like.

iii)   Do not use fire. The most fearful disaster in association with an earthquake is fire. In case you have to use fire, someone must be beside the fire. Otherwise you must depend on automatic extinguishing instruments.

iv)   Mind dangerous articles. Make sure that inflammables such as petroleum, benzine, edible oil and the like are put in a safe container. Are they well away from fire? Are propane cylinders fixed so as not to fall down? Are their valves closed?

v)   Prepare plenty of water in buckets and fire extinguishers in case of fire. Confirm how to operate the extinguishers. Fill bath tubs with water.

vi)   Listen to correct information. Keep TV and radio switched on all the time. Pay attention to the information from municipal offices, fire and police stations. Be cautious not to be affected by false information.

vii)   Put on light clothes suitable for work and evacuation in the worst case. Prepare hoods and helmets.

viii)   Check things which you must carry about with you in case of evacuation. Have you prepared drinking water, food, a portable radio, a flashlight, medicine and so on?

ix) Confirm the evacuation place. Secure the exit from your house. Do you know the evacuation place? Have you confirmed the roads leading to the evacuation place? In an area under the threat of tsunami or landslide, follow the instructions by police and fire-brigade.

(x) Help each other with your neighbors. Consult beforehand with neighbors about fire fighting and evacuation.

(2) Ten key-points at places of work.

i) Have a meeting of members in your working place under the direction of the person in charge of disaster prevention. Take all possible measures by assigning an appropriate job to each member.

ii) Inspect your working place and find the safest place. Confirm the place where you stay for the time being in case of earthquake occurrence. Have you fixed heavy machines, lockers and the like in such a way that they will not fall down? Attach tapes on windows, so that broken pieces of glass do not scatter.

iii) Limit use of fire as much as you can. Take necessary measures for safety of your equipment, machines and the like.

iv) Check dangerous articles and places according to the existing short-term plan for disaster prevention.

v) Self-defense fire-brigade should stand by. Check anti-fire instruments, water for fire fighting, independent power plants and so on.

vi) Check things such as important documents, negotiable papers and the like to be carried away in case of emergency.

vii) Stay at a safe place. When your working place is built strong enough for ground vibration and fire, you may stay there. If not, you had better evacuate. Decide immediately what you should do and take action at once.

viii) In department stores, theaters and the like where there are a large number of visitors, safety of these people is the matter of the highest priority. When evacuating them, proper guidance and direction should be made so as not to cause a tumult.

ix) Obtain correct information. Watch and listen to TV-radio at all times. Pay attention to information from prefectural and municipal offices, fire and police stations.

x)   Help each other with people in neighboring work places.
It is important for neighboring enterprises to cooperate with one
another in the case of fire fighting and rescue operation. Such
cooperation is especially important for those working in over-
populated areas or in buildings in which people of various types of
employment are working.

11.5.15   Do not use a car for evacuation

If everyone tried to run away by car, it is obvious that all roads
would be jammed by traffic, so that no cars would move at all.
Moreover, it becomes difficult for emergency vehicles for fire fighting
and rescue operations to move. Cars filling the roads may readily be
burnt as they are carrying gasoline.

When you are informed of the result of Prediction Council while
driving a car, please obey the instructions of local police.

What have been written in this section comprise provisional model manuscripts which would be put on the air by NHK when the
Prediction Council is called and when it concludes the possible
occurrence of the Tokai earthquake.

Such manuscripts will certainly be modified as the earthquake
countermeasures are improved in the future. The emergency broadcasting is a matter of deep concern not only for NHK but also for
commercial broadcasting corporations in the earthquake areas.
Those media using broadcasting techniques should investigate in
detail how to cope with earthquake prediction information.

Newspapers, weekly and monthly magazines are also requested
to deal with earthquake prediction information, or earthquake
warnings statement and other associated matters in a responsible
way. The media are in a position to advise and appeal to citizens
regarding the correct way to fight against the coming earthquake.
They can teach the mass how to behave in case an earthquake
warning is announced.

CHAPTER 12

PUBLIC REACTIONS AGAINST
AN EARTHQUAKE WARNING

Earthquake prediction has become a matter of actuality since the later half of the 1970's. As earthquake prediction programs in Japan, China, the U.S.A., the U.S.S.R., New Zealand, and other countries have developed, the nature of earthquake precursors has been gradually brought to light. Accordingly, it has been supposed for some time that occurrence of an earthquake might, in some cases, be predicted to a certain extent. Actually, the first success of prediction was the Haicheng earthquake ($M = 7.3$) on February 4, 1975 which was reported from China. This was really the first time for a destructive earthquake forewarning and many lives were saved by the issuance of such an imminent warning of earthquake occurrence.

Debate about the difficulties in converting earthquake prediction information into an earthquake warning has at the same time arisen among sociologists and seismologists even though earthquake prediction is actually achieved. This is especially so in highly developed, capitalistic countries like the U.S.A. and Japan. It was thought that extraordinary social unrest would be induced if a warning of occurrence of a destructive earthquake having a magnitude 7 or larger should be abruptly issued from an accountable organization over a densely populated area.

On the occasions of the 1965–66 Matsushiro earthquakes (see Section 8.1) and the anomalous uplift in Kawasaki City, Japan in 1974 (see Section 4.1), we observed some tumult among local inhabitants although the information communicated to the locale was hardly called a warning of occurrence of a highly hazardous earthquake. The magnitude of suspected earthquake was estimated as 6 or thereabout at maximum in these cases and the local people had plenty of time to prepare for the expected earthquake. Therefore, no public panic took place.

As was stated in Subsection 10.2.1(1), however, the Ishibashi hypothesis, which was reported by mass media in a highly strained emotional atmosphere shocked and concerned many of the local citizens. Even if it cannot be said that they were really panicked, the point that a mere theory or hypothesis of a young scientist, about occurrence of a great earthquake, affected the locale to such an extent seems important. Should the time, location and magnitude of a great earthquake be officially foretold by an accountable government organization, it is apparent that some sort of panic would take place. That is especially so, should the prediction information be suddenly released without any precaution. Such a situation occurred to some extent by the public reaction against the aftershock information issued by the Shizuoka Prefectural Office as discussed in fair detail in Subsection 11.4.2.

In Japan, the situation has been dramatically improved in recent years because of the enactment of the Large-scale Earthquake Countermeasures Act in 1978. However, there are many difficult points to be solved in relation to issuance of an earthquake warning.

Let us assume that a medium- and short-term earthquake prediction is made based on precursors of the first kind. When the magnitude of coming earthquake is equal to or larger than 7, the lead time would in general be several years or longer. In such a case, the information would soon leak to the mass media so well organised as active in a country such as Japan. It is a difficult problem for administrators responsible for disaster prevention to convert such information into an earthquake warning. As the earthquake would not occur soon, even if it may occur at all, and as the estimate of occurrence time would involve an error ranging from several months to 1 year, it is extremely difficult to issue definite earthquake prediction information which is practically an earthquake warning.

Because of the possible leakage of information, local people in the area likely to be the target of the expected earthquake would become restless. Construction business would decline, and some inhabitants might think of running away from the threatened area. New investment enterprises would be doubtful. Some businesses might be closed. It is unavoidable, therefore, that the locale would face economic depression.

According to the Large-scale Earthquake Countermeasures Act, only long-term and imminent prediction information are taken into account. However, the author is of the opinion that medium- and short-term earthquake prediction information can be put forward by analyzing data taken by observation arrays. It is therefore necessary to consider what actions to take during the period of medium- to short-term predictions.

At any rate, we do not know how inhabitants, private enterprises, public corporations and the like in the locale would take measures against coming disaster as earthquake prediction information became more accurate, month by month and day by day.

Summarizing the author's statements in this chapter, it is highly important to know reactions of the public against earthquake prediction information or earthquake warnings of various time and intensity levels. If administrators can correctly judge such reactions, they would be able to adjust timing and content earthquake warnings, which would become more accurate and specific as time goes on, in such a way that very little social panic ensues.

12.1 Earthquake Prediction and Public Policy—A Problem Brought forward by the U.S. National Academy of Sciences

The U.S. National Academy of Sciences was the first body to take up the problem concerning earthquake prediction and public reaction. A panel consisting of not only seismologists but also of sociologists was formed in the Academy, and a report entitled "Earthquake Prediction and Public Policy" (NATIONAL ACADEMY OF SCIENCES, 1975) was published. That contained in the report is not the result of actual researches. Only important points relevant to the problem are stressed in the report. But the author thinks that mentioned indicates almost all the important problems for converting earthquake prediction information, which is a purely scientific matter, into an earthquake warning that involves much of socioeconomic problems.

There is no room in this book to cover all the contents of the report. But the author believes that it is useful to quote the conclusions of the report, which were stated as recommendations, in the following.

Recommendation 1 The highest priority in responding to earthquake prediction should be assigned to saving lives, with secondary attention to minimizing social and economic disruption and property loss, provided the costs of specific measures are within the limits that society is willing to accept.

Recommendation 2 Prediction should be used in conjunction with a complete program of earthquake-hazard reduction, and not as a substitute for any of the procedures in current use.

Recommendation 3 The primary responsibility for planning and responding to earthquake predictions should be assigned to federal, state, local, and private agencies having broad concern for community and economic planning and for disaster preparedness and response, rather than to newly formed agencies established especially to deal with earthquake prediction and warning or to agencies concerned primarily with emergency response.

Recommendation 4 As an essential feature of advance planning, legal determinations and clarifying legislation should be sought to minimize the legal ambiguities that otherwise will hamper officials in making constructive response to earthquake prediction.

Recommendation 5 Legal inquiry should be undertaken to clarify what powers for responding to earthquake predictions now exist under the Disaster Relief Act of 1974 (PL 93-288) and what further powers might be necessary. Any deficiency or uncertainty regarding application to the emergency created by prediction of a potentially destructive earthquake should be promptly corrected by new legislation.

Recommendation 6 A public agency should be assigned the responsibility of (a) identifying groups of people most likely to need special assistance in the event of an earthquake or to suffer disproportionate loss and disruption when an earthquake is predicted, (b) developing a plan to offset, insofar as is practicable, the inequitable costs and suffering attendant on both the quake and the prediction, (c) monitoring events after the prediction from the point of view of equity, and (d) helping unorganized population segments to recognize how the earthquake prediction affects their interests.

Recommendation 7   Predictions should be developed, assessed, and issued to the public by scientists rather than by political officials. Procedures must be developed to ensure the free and timely flow of information concerning predictions to all segments of the public. Legislation may be required to assure that information that an earthquake will occur at a given location and time will not be withheld from general knowledge to the advantage of special interests.

Recommendation 8   A designated federal agency should establish a group of governmental and nongovernmental scientists who can be called upon to evaluate specific earthquake predictions. The responsibility for establishing this group should not be vested in any agency that is involved in the technical pursuit of earthquake prediction. This agency should also maintain a public record of all published predictions.

Recommendation 9   A designated federal agency should confer promptly with governors of the principal earthquake-prone states or their representatives to clarify the respective responsibilities at each level of government and to establish procedures for issuing earthquake warnings.

Recommendation 10   A warning should be issued by elected officials promptly after a credible prediction of a potentially destructive earthquake has been authenticated. A warning should include a frank assessment of the prediction, noting the possibilities for error, information on the types and extent of damage that the earthquake could cause, a statement concerning plans being developed to prepare for the quake, and advice concerning appropriate action to be taken by individuals and organizations.

Recommendation 11   A designated federal agency should establish mechanisms for monitoring public understanding credence, and response at all stages of the prediction-warning-earthquake sequence, and for making this information available promptly to responsible public officials.

Recommendation 12 Careful attention should be paid to the problems of communicating to segments of the population that might otherwise receive only last-minute warnings. These segments include such groups as foreign-speaking minorities, the physically handicapped, tourists, and the socially isolated.

Recommendation 13 As part of a complete and continuing earthquake-mitigation program, each earthquake-vulnerable community should develop a hazard-reduction program, involving both public and private agencies, to be put into effect in case of an earthquake warning. A designated federal agency should establish a central clearinghouse to provide the necessary hazard-reduction information and technical assistance to states, which in turn will aid communities in developing their plans and in implementing them.

Recommendation 14 Each threatened community should examine the applicability of each of the following major kinds of hazard-reduction measures; (a) evacuating limited areas and vacating dangerous structures; (b) accelerating structural design and maintenance programs; (c) employing land-use planning and management powers in relation to the predicted locale of the quake; (d) protecting essential natural gas lines and other community lifelines; (e) dealing with such possible hazards as nuclear plants, vulnerable dams, highly flammable structures and natural cover, and facilities involving the risk of explosion or the release of dangerous chemicals.

Recommendation 15 It should be accepted policy on the part of public and private agencies that a considerable part of the financial assistance normally available to a community after an earthquake should be made available as needed for hazard-reduction measures taken in response to an authenticated prediction of a potentially destructive earthquake. New legislation should be enacted as required to achieve this end, taking into account the example of such existing legislation as PL 93-288, the Disaster Relief Act of 1974, especially Title IV, Section 417 of that Act on "Fire Suppression Grants" (see Recommendation 5).

Recommendation 16 Emergency plans in earthquake-vulnerable areas should be revised to include programs for readying emergency services in the interval between warning and quake.

Recommendation 17   Emergency plans should include programs for broad and active citizen involvement in preparing for the earthquake.

Recommendation 18   Upon issuance of an earthquake warning, a joint governmental and private-sector commission should be established to monitor the economy in the threatened area to ensure early detection of changes and to make recommendations to government, business, and labor organizations as needed. Representatives of insurance and investment organizations should be included and should play an integral part in the work of the commission.

Recommendation 19   In the event of a credible earthquake prediction, policy makers must continously weigh the relative merits of sustaining the economy in the threatened area at its prewarning level or of encouraging some orderly outflow of capital. Economic subsidies may be required either to sustain the economy or to protect groups of people who would otherwise suffer undue hardship as a consequence of economic dislocation resulting from the prediction and warning.

Recommendation 20   Consideration should be given to the development of standards to govern the practices of businesses and individuals offering services related to earthquake mitigation to the public.

The recommendations in the above are so reasonable that they can be applied to the problems concerned even in Japan. The key points of these recommendations are as follows;

i)   To give the highest priority to saving lives.

ii) To make necessary legislation.

iii) To open all information to the public.

iv) To set up an earthquake prediction evaluation committee.

v)   To make clear who takes the responsibility of an earthquake warning.

vi) To take measures for protecting minorities and/or tourists.

It is also pointed out in the report that various researches should be conducted in order to achieve these recommendations.

They are indicated as "research recommendations" amounting to 14 in number. Stress is especially put on the socioeconomic influence of an earthquake warning and the legislation that makes issuance of an earthquake warning possible.

12.2 Study on Public Reaction against an Earthquake Warning Made by a Research Group at the University of Colorado

A sociological group at the University of Colorado (HAAS *et al.*, 1975; HAAS and MILETI, 1976, 1977a, b) carried out studies on probable public reaction against an earthquake warning at Californian cities which are under potential threat of large earthquakes. The reactions are studied through interviews and enquiries with important members of city councils, polices, fire stations, hospitals, water service, sewer and road departments, gas, electricity and telephone corporations, production factories, retail dealers, real estates, banks, developers and so on. The so-called "Delphi method" was used for the investigation. As a result, some idea about what various organs in the locale would do in case of an earthquake warning could be obtained. Such studies were also extended to individuals.

The author takes the liberty of quoting the scenarios used for the studies by the Colorado group (J. E. Haas, personal communication, 1975) in the following two subsections.

12.2.1 Illustrative mini-scenario for an official at the local organization level

There is an earthquake forecast developed by one of the most highly respected universities in this state. The Governor's office asked an independent committee of experts to carefully examine all of the evidence to see if the forecast is justified. The professors who developed the forecast originally have cooperated fully with the committee of experts.

The committee has concluded after careful study that there is an adequate scientific basis for the forecast. The forecast elements are as follows:

An earthquake of Richter magnitude 6.6 (comparable to the 1971 San Fernando earthquake) will occur in approximately 36 months from now (plus or minus one month). The epicenter as computed will be within an area already identified. (See the red arrow on the attached map.)

The principal radio, television, and newspaper organizations have indicated that they will be giving the forecast prompt, full and complete coverage. Indeed they say that the coverage will be comparable to the first U.S. man on the moon event.

The seismologists and earthquake engineers to whom we have spoken indicate that there is likely to be very heavy damage to buildings in about 20% of your city, moderate damage for an additional 50%, and minor damage for the remainder. News reports on these damage estimates are certain to come out in the near future.

Federal agency officials say that under the current law no massive unusual Federal assistance is possible. Normal programs can possibly be speeded up, however.

Your two U.S. Senators have introduced a bill which would provide some emergency financial help but it is not clear what would be the final form of any such bill if indeed it passes at all.

Some state legislative committees have begun discussing the state's responsibility in this unusual circumstance, but what comes of that seems impossible to predict now.

All of the insurance companies have agreed to suspend the sale of earthquake insurance on all property within a 50 mile radius of the forecasted epicenter.

The large banks and investment firms around the state are being very close-mouthed about their plans, except to say that, as in the past, they must invest their resources with an eye to minimizing risk and maximizing return. The *Wall Street Journal* says that it expects commercial mortgage money for this area to dry up within six months unless the forecast is refuted by seismologists at the U.S. Geological Survey in an official pronouncement.

The California Department of Employment is anticipating a modest decline (5–7%) in employment in this area in the next 2 years.

The Governor's Office of Emergency Services is prepared to offer technical assistance in any disaster preparedness programs which the city may initiate.

[Note: The remainder of the scenario would be filled in with summary statements based on what we had learned from officials at the *local level* during our informal discussions with them.]

12.2.2   Illustrative mini-scenario for an individual community resident

After a very careful scientific appraisal and evaluation, a forecast has been made by experts that an earthquake of Richter magnitude 6.6 (comparable to the 1971 San Fernando earthquake) will occur in about 36 months from now (plus or minus one month). The center of the earthquake will be somewhere within the city limits of your town. Scientists say that 20% of the town will have heavy damage, 50% will have moderate damage, and the rest will have some minor damage.

Most of the radio, television and newspapers will give the forecast extensive coverage and make these damage estimates public very soon. What the Federal and state governments will do to help is still unclear.

Insurance companies have agreed to suspended the sale of all earthquake insurance in your area. The large banks and investment companies in the state (and those in your area) are closed-mouth about their plans. They have said that they will invest their money with an eye to minimizing risk to them. The *Wall Street Journal* expects mortgage money for the area to dry up within 6 months. Builders plan on finishing the buildings and houses already under construction now but aren't planning on starting anything new. Most realtors in the city plan on encouraging their local clientele to look toward the surrounding counties for any moves they might have in mind.

Unemployment in your area is expected to increase. The branches of the major department stores in the area are considering a reduction in their personnel by converting to a self-service system.

The Governor's Office of Emergency Services is prepared to give your city technical assistance in getting ready. But the way things look now this will increase city taxes soon by about 6–7% to pay for programs such as strengthening old public buildings, planning for the coordination of inter-county fire department services, and getting ready to bus students in the city schools to schools in surrounding areas as the time approaches.

The local Red Cross, Salvation Army and hospitals here and in other counties are revising their emergency plans.

Most local businessmen say they'll continue business as usual for now, but plan to close down as the time approaches (either for long vacations or to do things to protect their investments).

[Note: Some illustrative follow-on lead questions asked of individuals may include: (A) Do you anticipate a vacation out of the area about the time of the forecasted earthquake? (B) Would you object to your children's being bused to nonlocal school? and (C) Would you make any structural changes to your house to try to cut down on possible loss?

12.2.3   Brief summary of the results of investigation

The investigation by the Colorado group was carried out for two cases in which it is assumed that earthquakes on magnitude 7.3 and 6.3 are predicted by an accountable organizations, respectively.

In the following, the results of investigation for the magnitude 7.3 case will be briefly summarized. As can be seen in the block-diagram in Fig. 12.1, the local economic activity seems likely to be affected very much.

It is assumed that precursors for an earthquake of this magnitude

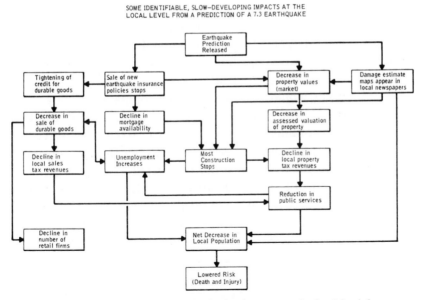

SOME IDENTIFIABLE, SLOW–DEVELOPING IMPACTS AT THE LOCAL LEVEL FROM A PREDICTION OF A 7.3 EARTHQUAKE

FIG. 12.1.   Some identifiable, slow-developing impacts at the local level from a prediction of a 7.3 earthquake (HAAS and MILETI, 1976, 1977a).

would be detected about 3 years prior to the earthquake occurrence at the latest and that the locale would be nominated to an area of intensified observation. The time-window of earthquake occurrence can be estimated as 2 months with a probability of 50% 2 years prior to the earthquake occurrence. One year before the occurrence, the time-window would be assessed as 1 month with a probability of 80%. It is later expected that the accuracy of estimating the time-window would become 1 week. It seems to the author, however, that the assumptions for estimating the time-windows are too optimistic at the present stage of earthquake prediction study.

When the first warning is made public, the mass media would report it as a front-page story. Much debate would arise about the reliability of earthquake prediction.

The administrative authorities would make a thorough investigation of expected damage, so that a map of disaster distribution would be reported in newspapers.

Safety of dams and reservoirs would be investigated by the owners. When the expected time of earthquake occurrence is approached, more measures for safety would be taken, for instance, by lowering the level of water head. The owners of skyscrapers would begin to check their safety, so that experts of earthquake-proof construction would become much in demand. Communication and computer systems, which are not earthquake-proof, would be specially reinforced or moved to a safe place.

Local citizens would rush to insurance companies to buy earthquake insurance. As the needs would exceed the capacity of insurance companies, the commissioner of insurance would be forced to suspend new earthquake insurance. After all, earthquake insurance subsidized by the government would be created.

Value of all properties in the area in question would become as low as 60% compared to that before the warning. The number of people, who would like to sell their property, would be far larger than those who want to buy. But there would be some people who would buy land very cheap at this time and sell after the quake for a high profit. Such an action would stop the extreme fall of land price. The buying of houses with mortgages would be stopped. People who bought a house with a mortgage selling scheme would tend not to pay the monthly fee because they know that the constructors cannot sell them, even if they repossessed the houses.

During the 2 years' period forerunning the earthquake, the population would decrease by 10%. During the last several weeks, 60% of the remaining population would voluntarily run away. Some large enterprises would leave the area concerned forever. No new enterprises would come into the area.

Construction businesses would entirely be suspended in the area. Banks and credit associations would become cautious not to be faced with a run on them. Credit purchase would become limited. Small businesses would meet difficulties, so that they would try to unite with larger businesses. During the period that corresponds to the time-window of earthquake occurrence, all the businesses would completely shut down and the people engaged in the businesses would be evacuated temporarily. In summary, economic activity would become lower by 25%. Most enterprises would try to decrease their stocks in order to lessen the damage in case of earthquake.

In the case of the prediction of a magnitude 6.3 earthquake, the above response of the public would become somewhat different. As the lead time from the first warning to the earthquake occurrence is assumed to amount to only 9 months, it would be difficult to take adaquate countermeasures in time.

The public reaction against an earthquake warning as mentioned

TABLE 12.1.   Condensed comparative findings on the use of earthquake prediction information (HASS et al., 1975).

		7.3 Richter magnitude prediction	6.3 Richter magnitude prediction
A.	Specificity and certainty	Three years lead time, two years as an official prediction; probability range from an initial 25% to 80%; time window range from two months to seven days	Nine months lead time, five months as an actual prediction; probability of 50%; time window range from one month to ten days
B.	News media coverage	Intensive coverage, then spaced over time, intense again as date nears; full range of opinions, actions, and consequences covered	Same as for 7.3 magnitude, but spaced differently because of shorter prediction period

TABLE 12.1 (continued)

		7.3 Richter magnitude prediction	6.3 Richter magnitude prediction
C.	Damage estimates	Extensive damage assessments by government agencies; maps of expected damage presented in media	General damage assessments; maps of expected damage presented in media
D.	Mitigation measures	Analysis of seismic vulnerability of dams, reservoirs, and large buildings; some risks lessened; major vulnerable systems moved or protected	Minor safety precautions taken in some business organizations; vulnerable dams drawn down
E.	Insurance	New earthquake insurance not available; governmentally subsidized insurance available just prior to earthquake	Earthquake insurance (new policies) prohibited; governmentally subsidized insurance not made available
F.	Property values	Drop to 60% of pre-prediction values; land speculation; mortgage money unavailable; some foreclosures	Drop to 80% of pre-prediction values; higher down payments required for mortgages
G.	Population and business movements	Net decrease of 10% of population over last 2 years of prediction; 60% voluntary evacuation during last few weeks; some large businesses move out permanently; some organizations move temporarily	Voluntary evacuation of 20% of population in last 3 weeks of prediction; businesses do not relocate, but do take safety precautions
H.	Changes in business activity	Construction halts; restrictions on credit; some foreclosures and bankruptcies; reduction of inventories; many shift business activities to safer areas temporarily; 25% overall decline in business	Few changes in normal activities; postponed new construction; varied precautionary and safety measures taken for the predicted time window; several close for same time period; leaves granted to workers

above can be summarized as shown in Table 12.1.

12.2.4 Comparison between the U.S.A. and Japan

The Colorado group came over to Japan in 1975. With cooperation of Professors T. Nakano and K. Abe, they examined the public reaction against the pseudo-warning of an earthquake at Kawasaki City in 1974 (see Section 4.1) through interviews with responsible persons of CCEP, Tokyo Prefectural Office, Kawasaki City Office, police and fire stations, railways, production factories, insurance companies, department stores, and so on.

It is interesting to note that they found remarkable differences in the reaction against an earthquake warning between the U.S.A. and Japan. The Colorado group was much impressed by the fact that no real panic occurred in Kawasaki City and that local citizens seem likely to appreciate the information notwithstanding it turned out to be false after all. Most local people with the exception of a few who are depending on construction, retail estate businesses believe that a large earthquake would hit the locale sooner or later, so that they think that it is a good thing to intensify earthquake countermeasures.

The Colorado group was also shocked by the fact that very few people pay attention to earthquake insurance in Japan. The upper limit of insurance is 2,400,000 yen for an ordinary house. The insurance on the house can be paid only when it is completely destroyed. The amount is one order of magnitude lower than that needed for building a new house. The insurance rate is also extremely high. These are the reasons that Japanese people are not interested in earthquake insurance. In 1980, the upper limit for an ordinary house was increased to 10,000,000 yen.

It was made clear, however, that Japanese insurance companies would not sell earthquake insurance any more in case an official earthquake warning is issued. This is also the case for American insurance companies.

Public reaction against an earthquake warning could be different from nation to nation depending upon national traits, economic condition, size of the country and so on. The author should here like to emphasize the importance of studies on public reaction against an earthquake warning. It is regrettable that no intensive study into these problems has been undertaken in Japan.

In Table 12.2 is shown the comparison of public reaction against an earthquake warning as summarized by the Colorado group (J. E. Haas, personal communication, 1975). It is seen in the table that

TABLE 12.2. Comparison of public reaction against an earthquake warning between Japan and the U.S.A. (J. E. Haas, personal communication, 1975).

Item	Japan	U.S.A.
Fear complex of a large earthquake	Very much feared.	Very few people know what a large earthquake is like even in California.
Earthquake insurance	Very little concern.	Rush to buy insurance in case of an earthquake warning.
Attitude of businessmen and officials	No remarkable action such as American one would be taken.	Head of an organization tries to increase budget and membership taking advantage of the warning.
	No buying up of large scale would occur.	Realtors would try to buy lands very cheap before the earthquake. They want to sell the lands with a higher price after the shock.

American businessmen want to make money even when local people suffer from disasters.

12.2.5 Development of public reactions against earthquake prediction as the information becomes precise with time

HASS and MILETI (1976) traced possible organizational responses on the basis of a scenario in which they took the point that earthquake prediction information becomes accurate as time goes on. The outline of their results will be reproduced in a summarized form below.

The following 4 stages are assumed for earthquake prediction information in their scenario.

Stage 1 (38 to 25 months prior to the earthquake)

USGS detected a precursor-like anomaly near Los Angeles. Two reputable seismologists predicted a damaging earthquake along the San Andreas fault in 3 years (25% probable). But this is not an official announcement of earthquake prediction.

Stage 2 (25 to 10 months prior to the earthquake)

An official prediction was announced by USGS about 25 months prior to the earthquake. An $M \geq 7$ earthquake was predicted with a probability of 50%.

Stage 3 (10 to 3 months prior to the earthquake)

USGS revised the prediction 9.5 months prior to the earthquake in such a way that an earthquake of magnitude 7.1 to 7.4 would occur with a probability of 80%.

Stage 4 (3 to 1 months prior to the earthquake)

USGS refined the earthquake prediction further about 3 months before the earthquake. The quake of estimated magnitude 7.3 would occur some time during the first week of the month specified. The probability was still 80%.

Public responses for the above stages are respectively investigated and are summarized as follows;

Responses for stage 1

1. Newspapers immediately reported that "intensive study" meant scientists would be placing more instruments in the area to better measure what was happening in the earth.

2. County and city officials assured the public that there was little cause for worry. Reports in the newspapers, however, made it clear that there was concern for the situation—there were increasing contacts between local and state officials responsible for public works and construction about what state agencies would do if the situation became more serious.

3. Newspapers began series on earthquake prediction, explaining that several successful predictions had occurred before; in China many lives had been saved by the prediction of a large earthquake, and in the USA some small earthquakes had been predicted. However, it was noted that some predictions had been wrong. The newspapers also quoted economic experts as saying that an official earthquake prediction would hurt the economy of the city.

4. Eleven months later, it was reported that population growth in the area had slowed. Fewer new businesses had opened, and most construction had stopped.

Responses for stage 2

1. Immediately after the official prediction, the Governor announced that his Advisory Panel on Earthquake Prediction had examined the Federal government data, and agreed with the prediction. The Governor directed state agencies to prepare for the possible earthquake. Federal agencies were doing the same.

2. Although some local government officials doubted the prediction, nearly all said that appropriate actions would be taken by the city and county departments involved.

3. Within a few days of the prediction, news reports showed that the majority of scientists found the prediction believable. A few scientists said, however, that predictions were not yet possible.

4. Shortly after the prediction, the State Insurance Commissioner announced that new earthquake insurance policies would no longer be sold to people living in the area, although people who already had earthquake insurance could keep it. Insurance companies and mortgage lenders began to call upon the government to provide new earthquake insurance programs to help prevent property values from dropping.

5. Within a month after the prediction, most newspapers had published maps of the area which showed where the damage would be.

6. Engineering firms in California began to advertise that they would inspect buildings and make recommendations for making them safer. Some business firms began to hire engineers to examine their buildings, and government agencies began to inspect public buildings and dams for safety in case the earthquake occurred. Some government agencies prepared pamphlets on safety measures citizens should take.

7. Planning for construction within 25 miles of the predicted center of the earthquake stopped immediately. Construction work already in progress was continued. Because of the slow-down in construction, unemployment in the building trades had reached 80% 3 months after the predictions; people who worked in these trades were looking for work elsewhere.

8. Savings and loan associations operating in the area were forced to reduce, by as much as 80%, the loans they could make in the area, and mortgages were less available. Since new loans required higher down payments, fewer people could get loans and property became difficult to sell.

9. Seven months after the "official prediction", there was a slight decrease in the money collected from sales taxes. It began to look like some services provided by the government would need to be

cut. Long-term planning for the city was drastically revised.

10.   One year after the "official prediction", the U.S. Congress began to consider a Federal insurance plan for the predicted earthquake area, because of the economic problems there.

Responses for stage 3

1.   On the heels of the revised prediction, media reports showed that most local officials found the prediction believable, and that they were calling for action in preparation for the earthquake from appropriate agencies on all governmental levels. After learning that the state would have inadequate funds, the Governor announced that California was seeking a Presidential Emergency Declaration to provide money for special preparedness measures, and to cope with the severe economic problems in the threatened area. Many employers, both public and private, began to urge their employees to plan their vacation period to coincide with the expected earthquake.

2.   Local and state governments speeded up their planning for response before and after the predicted earthquake. Public information drives to familiarize citizens with measures were intensified.

3.   Measures were taken to reduce the danger from dams and fuel lines that might break in the event of the earthquake. After inspection by state experts, there were plans to partially empty unsafe reservoirs, by that time—2 years after the "official prediction".

4.   A few months before the expected date of earthquake, work was begun to move some governmental offices and equipment out of the area temporarily. Some businesses began to move vital records and sensitive equipment; other businesses began to institute measures to protect their stock. Some large industries announced plans to close down during the time of the predicted earthquake to protect employees. Schools also announced that their opening in the fall would be delayed. A few national business firms in the area moved their facilities and operations to other locations. Because of increasing economic pressures, many small businesses in the area closed their doors. About 8 months prior to the expected date, 25% of them had either declared bankruptcy or sold out.

5.   Due to decreased business activity, many financial firms and larger retail firms were forced to lay off employees. This was a blow to the local economy. Local businesses and Chamber of Commerce

officials began campaigns to assure the public that, in spite of some economic problems, the area and its businesses were basically safe and sound.

6. By the time about 3 months before the expected occurrence of the shock, it was apparent that the income to local government from taxes had declined.

7. City and county departments concerned with vital services, such as police, sheriff, fire and water departments, requested extra money in order to get ready for the earthquake. The increases could not be granted, but sufficient money for vital services was obtained by cutting funds to libraries, parks and recreation, trash collection, and street cleaning.

Responses for stage 4

1. All efforts to prepare for the earthquake were speeded up. Local government officials urged evacuation of high-risk areas, particularly of areas below some reservoirs. Officials said that police and fire protection for empty houses and buildings would be stepped up, although total protection could not be guaranteed.

2. A Presidential Emergency Declaration for the area was finally announced, and officials and businesses hurried to get funds forthcoming from the declaration. The funds were used to further prepare for the possible emergency and to lessen some pressing economic problems in the area.

3. Shelters were established by the Red Cross and other agencies to handle residents who evacuated before the possible earthquake.

4. Property values had dropped to their lowest point in the last three years, and even lower in areas of expected high damage.

5. General unemployment had risen markedly; unemployment in the construction industry had been high for a long time.

6. Some supermarkets and other retail stores announced they would stay open during the week of the predicted impact.

7. At the direction of the Governor, the National Guard was sent into the area to assist local authorities.

8. Hospitals and prisons transferred their charges to safer locations.

9.   Some overpasses and bridges were closed and detours through safer areas were established.

10.   As the outflow of population accelerated during the last two weeks, many businesses closed down entirely and the buildings were vacated.

11.   One week in advance of the earthquake week, all public buildings were vacated. Records had already been given special protection or relocated. For the few critical governmental functions that continued, skeleton crews operated out of trailer-type offices located well away from buildings and power lines.

The author takes the liberty of quoting the above public responses for respective stages of earthquake prediction information almost as they are in HAAS and MILETI (1976). It appears to the author that such probable public reactions as revealed by the study would give some clue to estimating reactions in case of actual issuance of earthquake warning.

The chain of public responses in the above can be illustrated in a block-diagram as shown in Fig. 12.1. As a result of the decrease in local population that follows the decreases in property values and construction businesses, death and injury would considerably decrease.

At any rate, it becomes apparent that almost no lives would be lost if an earthquake warning could be issued and local people would take it seriously. It is possible, however, that the economic loss during the period from the issuance of warning to the earthquake occurrence could be so large that the loss is larger than that caused by the quake itself in an extreme case. It is highly important, therefore, to choose the best timing for issuing an earthquake warning in order to minimize economic losses.

12.3   Case Study in Kawasaki City in Connection with the Anomalous Uplift

As was stated in Section 4.1, an anomalous uplift centering at Kawasaki City adjacent to Tokyo (GEOGRAPHICAL SURVEY INSTITUTE, 1975) was detected in 1974. It had one time been feared that such an anomaly might have something to do with an earthquake occurrence there although, in fact, no earthquake occurred after all.

Oнтa and Abe (1977) interviewed many local citizens and investigated their reactions against the information although it was not really an earthquake warning issued officially by an accountable government organization. But it is clear that the Kawasaki case provided a good example for studying public reaction against an earthquake warning. The author can quote only some of interesting points from Oнтa and Abe (1977) in the following.

As a result of the investigation, it becomes apparent that many inhabitants believed that an earthquake causing ground tremors of intensity VI or VII in the JMA scale (VIII or IX and over in the modified Mercalli scale) would hit the Kawasaki area within 1 to 2 years' time. Even though the anomalous uplift could be associated with an earthquake, seismologists would have expected a little lower intensity and an occurrence time a little later. It is true, therefore, that information has been somewhat exaggerated when accepted by local people.

The percentages of people who thought that the earthquake would occur in various probability ranges are as follows;

Probability of 50%	21.3%
Lower than 50%	20.5%
Higher than 50%	21.1%
No answer	37.1%

In this case, it should be borne in mind that nothing had been announced about the probability of occurrence from an accountable organization such as CCEP.

About the feasibility or reliability of earthquake prediction, the percentages of local people are as follows;

Believe the prediction	51.1%
Do not believe the prediction	22.9%
No comments	23.2%

It is rather surprising that more than half of local inhabitants believe in earthquake prediction. This point should be seriously borne in mind by seismologists.

The percentages of people who took various measures against the expected earthquake are as follows;

Preparation of flash lights	53.4%
Arrangement of bank account books and	

other important documents in order   37.4%
Family discussions   48.1%
Preparation of transistor radios   35.0%
Storage of food and water   32.0%
Preparation of first aid supplies   29.5%

Such percentages of various preparations are higher for the aged generation than for the younger.

Those who think that earthquake prediction information is useful amounted to 69% of local inhabitants. Those who think it is useless amounts to 11%. About the timing for release of information, the following percentages were obtained by door-to-door enquiries:

When the slightest indication is found   38%
When the occurrence is 50% certain   26%
Only when it is quite probable   36%

They would like to have such information 6 months to 1 year before the earthquake with a percentage amounting to 75% and just 1 year before with a percentage of 21%.

It can scarcely be said that the Kawasaki case led to a panic. It is true, however, that economic activity was a little affected and that local citizens realized how poorly prepared the City Office was against earthquake disaster.

12.4   The "Aftershock Information" Associated with the Izu-Oshima Kinkai Earthquake and the Public Reaction

Much of communication problem and confusion in association with the "aftershock information" related to the Izu-Oshima Kinkai earthquake ($M = 7.0$, 1978) was discussed in Subsection 11.4.2. As the information was regarded as imminent earthquake prediction in many cities, towns and villages, we unexpectedly had a good example of public reaction against imminent earthquake prediction information in this case.

12.4.1   Actions taken by local inhabitants upon receiving the information

Table 12.3 is the results of the investigation performed by the Institute of Future Technology as published in SCIENCE AND

TABLE 12.3. Reactions against the "aftershock information" (SCIENCE AND TECHNOLOGY AGENCY, 1978).

Reaction*	Percentage**		Total number of people who took the action for the whole prefecture
	Shizuoka Prefecture as a whole	Izu district	
Tried to get further information through TV and radio	44.7%	53.0%	ca 1,000,000
Prepared things for emergency evacuation	17.5	33.2	390,000
Extinguished fire	15.4	31.7	350,000
Prepared drinking water	11.2	22.4	250,000
Consulted with family members, friends and neighbors	8.6	13.2	190,000
Went to buy bread and canned food	5.0	13.0	110,000
Thought that the information is false and asked other persons not to believe	3.5	7.2	80,000
Went home from working place	3.0	6.6	70,000
Went to school to meet children	1.2	2.0	30,000
Evacuated to nearby evacuation place	0.7	3.8	20,000
Took no action	23.6	12.6	530,000
Did not hear any such information	16.1	8.0	360,000

* Involves reactions against second hand information.
** Percentage who replied to the questionnaries.

TECHNOLOGY AGENCY (1978).

(1) Preparation for drinking water and food

About 110,000 people, 5% of the prefectural population, went to buy bread and canned food after receiving first- and secondhand information of aftershock. They also bought flashlights, candles, matches, helmets and so on. Most of the people who took these actions were living in the Izu district. This is certainly because of previous experience they suffered during considerable earthquake damage caused by the Izu-Oshima Kinkai earthquake and its aftershocks.

As can be seen in Table 11.7, city, town and village offices asked their citizens to prepare water and food, so that many people seemed to obey the instruction.

(2)   Fire extinguishing and other measures for dangerous articles

The number of people who prepared by extinguishing domestic fires amounts to 15.4% or about 350,000 throughout the Shizuoka Prefecture. The percentage was as high as 50% in the Izu district. The propaganda that the first task on the occasion of an earthquake is to extinguish fire is thoroughly popular among the Japanese. There are many people who secured propane gas cylinders, furniture and other items. But the percentage here is smaller than those who extinguished fire.

(3)   Looking for further information

More than 50% of people in the prefecture tried to gain information mainly through TV and radio when they were informed of the "aftershock information" including secondhand and/or false sources.

(4)   Commuting and other population movement

About 10% of people who received aftershock information when at their place of work or elsewhere returned home. In many instances, department managers ordered workers to go home upon receiving the information. Many schools and kindergartens took measures to ensure children went home earlier than usual. Many parents went to schools and kindergartens to collect their children. The percentages of those who took action mentioned here are indicated in Table 12.3.

(5)   Evacuation

In association with the "aftershock information", about 0.7% or 20,000 of the people in the prefecture moved to safe places. In turn, 25% people took refuge in the eastern area of Izu district where the damage associated with the main shock was large. About 18% (ca. 390,000) of the people in the prefecture prepared food, drink and clothes for evacuation.

The reaction of local people against the "aftershock information" as given in Table 12.3 and described briefly in the above can be seen more clearly in Figs. 12.2 and 12.3 (SCIENCE AND TECHNOLOGY AGENCY, 1978).

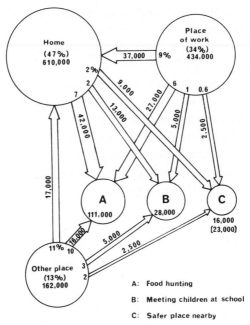

Fig. 12.2. Movements of local people after receiving the "aftershock information" as estimated for the whole Shizuoka Prefecture (SCIENCE AND TECHNOLOGY AGENCY, 1978).

12.4.2 Organizational response

It is interesting to see how various organizations in the Shizuoka Prefecture responded to the "aftershock information". Investigations by the Institute of Future Technology on such responses are summarized in SCIENCE AND TECHNOLOGY AGENCY (1978) although it is tedious and perhaps too much to quote all of them in this book. Those who are particularly interested are kindly requested to refer to the original report though the author points out that it is written in Japanese. Response of administrative organs at various levels, public corporations, schools, banks, hotels, and other private businesses are indicated in the report. It appears to the author that most organizations behaved very responsibly. They paid great detail to anti-fire measures, fixing propane gas cylinders and chemical containers on shelves, stopping the use of petroleum stoves, and so on. Woman workers were in general asked to go home early. However, male

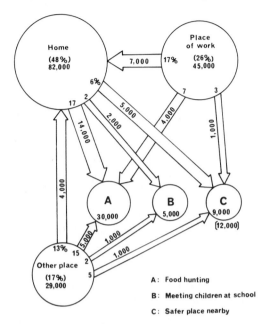

FIG. 12.3.   Movements of local people after receiving the "aftershock information" for the Izu district (SCIENCE AND TECHNOLOGY AGENCY, 1978).

employees at prefectural and municipal offices, police and fire stations, public corporations, propane gas dealers, and the like were asked to stand by even though they were off duty.

At department stores, staff members were specially instructed to prevent confusion among visitors, while those working in some hotels, pensions and souvenir shops were instructed not to divulge the information to visitors.

12.5   Problems Related to Swing-and-Miss Warning

That it is very problematic to issue imminent earthquake information or earthquake warning with present-day technology has been repeatedly stated in this book. Should a warning be issued and should no earthquake occur, it would be a matter of great difficulty to cope with social confusion and economic loss caused by such a swing-and-miss warning. Although it is expected with a high probability that swing-and-miss warnings would sometimes be inevitable, even if earthquake prediction technique develops to a level leading to the issuance of actual earthquake warning, nothing has as yet been decided about the measures to be taken at the time of a swing-and-miss warning.

The "aftershock information" associated with the Izu-Oshima Kinkai earthquake was not really a swing-and-miss warning because the Prefectural Office did not issue a warning. It released only information. However, cancellation announcement of the information had to be made public in order to quiet the tumult that arose following the aftershock information. As considerable people thought that the "aftershock information" was in fact an earthquake warning, the situation became much the same as that after a swing-and-miss warning. Subsection 12.5.1 will be devoted to describing how the prefectural and municipal offices and the mass media dealt with the situation. In Subsection 12.5.2 will be briefly stated the swing-and-miss warning in association with the Haicheng earthquake ($M = 7.3$, 1975) for which the main shock was successfully predicted. A fictitious documentary scenario in the case of a hypothetical earthquake prediction, which practically failed, will be quoted from WEISBECKER et al. (1977) in Subsection 12.5.3 in order to see what would happen in the U.S.A. in case of a swing-and-miss warning.

12.5.1 Tumult associated with the "aftershock information" of the Izu-Oshima Kinkai earthquake and cancellation announcement

As was mentioned in Subsection 11.4.2(1), the Shizuoka Prefectural Office released the "aftershock information" of the Izu-Oshima Kinkai earthquake at 13 : 40 on January 18, 1978. From about 14 : 00 hours in the afternoon, so many telephone enquiries about the information inundated the office that the telephone circuit did not function at all at about 14 : 30. This was also the case for city, town and village offices, police stations, weather stations and TV-radio broadcasting stations.

The Prefectural Office realizing that the "aftershock information" has been misunderstood and some confusion has been arising among inhabitants in the prefecture, put forward the Governer's announcement, that denied immediate occurrence of strong shocks, as indicated in Table 12.4 at 16 : 30. The announce-

TABLE 12.4. Governor's talk about cancellation of the deformed information as announced at 16:30 on Jan. 18 (SCIENCE AND TECHNOLOGY AGENCY, 1978).

I issued aftershock information at 13:30 today.

This was to inform especially the citizens living in the southern and middle areas of Izu district of the correct aftershock information and the measures taken by the Prefectural Office based on the conclusions of the Head Offce for Disaster Measures of the Izu-Oshima Kinkai earthquake of the central government. Accordingly, this information has to advise citizens not to be overly concerned and to behave calmly.

The Prefectural Office will take further necessary measures, so that I ask all the citizens in the prefecture not to worry unduly.

K. Yamamoto
General Director, Head
Office for the Disaster
Measures, Shizuoka Prefecture

TABLE 12.5. The broadcast of NHK Shizuoka station for cancelling the deformed information as put on the air during the Sumo (Japanese wrestling) program (SCIENCE AND TECHNOLOGY AGENCY, 1978).

The Shizuoka Prefectural Office released information about aftershocks this afternoon. There is some misunderstanding of the information, so that there is a rumor that a large shock will soon occur. This has caused some concern and further erroneous distortion of the information in some parts of the prefecture.

The Prefectural Office is asking citizens to clearly understand. The information was only to tell the citizens in the southern and middle areas of Izu district about the general possibility that aftershocks would further occur. Nothing about imminent danger was involved in the information.

The NHK will broadcast the news related to the aftershock information in detail including the official view of the Prefectural Office in the TV program "Shizuoka Today" at 6:40 this evening.

TABLE 12.6. Public relations announcement about cancellation of the deformed aftershock information in the Higashi-Izu Town as publicized by PR cars (SCIENCE AND TECHNOLOGY AGENCY, 1978).

This is to tell all the citizens.

An unfounded rumor that an earthquake will occur in 3 hours time is being circulated in the town right now. As this is absolute nonsense and has no scientific grounds, please behave calmly.

322

ment was sent to all municipal offices and other related organizations by the wireless communication facility for disaster prevention. Meanwhile, the Prefectural Office asked the mass media to cooperate by cancelling the distorted information that warned a large earthquake would soon occur or the ambiguous statements that led people to believe so.

From around 17:00 hours in the afternoon, municipal PR cars and wired broadcasting began to take action to cancel the deformed information. At the same time, TV and radio stations took similar action. Table 12.5 is what NHK broadcast suspending the regular program on the Sumo (Japanese wrestling), while Higashi-Izu Town broadcast the statements as shown in Table 12.6 by PR cars.

Around 6 to 7 o'clock in the evening, TV programs put detailed explanations about the "aftershock information" and the cancellation of deformed rumors on the air, so that all the confusion quieted down. This proved that the TV and radio media can be quickly influential for communicating the truth to the public. For the purpose of clarification and stopping the spread of false information, therefore, it is essential to make repeated broadcasts.

12.5.2 Swing-and-miss warning in association with the Haicheng earthquake

As already stated in Section 8.2, an order of evacuation was issued in the Panshan area, Liaoning Province on December 28, 1974, so that tens of thousands of people took refuge outside their houses in the freezing weather although no large earthquake occurred. The instruction for evacuation was issued only locally. The province revolutionary committee had nothing to do with that.

Nothing in detail about the reactions of local people against such a swing-and-miss warning has been obtained partly because of the social organizational system in China. The author gathered from Chinese colleagues, however, that people complained of the false warning.

12.5.3 Swing-and-miss warning in the San Francisco area—A fictitious documentary—

No actual example of public response to an earthquake warning is possible in the U.S.A. because the seismicity is lower than that in Japan and China. WEISBECKER *et al.* (1977) summarized the results of their study on public response to an earthquake warning in the form of a fictitious report of hearings held after an earthquake in an area near San Francisco. It is assumed that an imminent earthquake prediction was issued over the San Francisco area and that a magnitude 6.5 earthquake actually occurred as predicted in the San Juan Bautista area without causing any damage in San Francisco.

The warning was not entirely false because the earthquake actually occurred. But it is almost a swing-and-miss one for citizens in San Francisco. The author thinks, therefore, that the documentary would provide a good example of public reactions when a swing-and-miss warning is issued over a densely populated city. Everything in the scenario is fictitious, including all dates and names of persons involved.

The circumstantial setting is given in Table 12.7. The documentary takes a form of a report written by a science writer called G. M. Dearbone. The author quotes the whole documentary from WEISBECKER *et al.* (1977) although it is somewhat lengthy.

TABLE 12.7.   Circumstantial setting of the fictitious earthquake prediction in an area near San Francisco (WEISBECKER *et al.*, 1977).

Date	Event
Feb. 13, 1979	A prediction is made by use of the partially completed northern California USGS earthquake-prediction network that an earthquake with a magnitude of 6.5 to 8.3 will occur, with its epicenter lying between San Juan Bautista and Bodega Bay. The prediction is that the event will occur with an 85 percent probability within the next four days.
Feb. 15, 1979	An earthquake of magnitude 6.5 occurs within its epicenter at San Juan Bautista at 9:05 am PST.

HEARINGS ON SAN FRANCISCO EARTHQUAKE PREDICTION
Good Decision—Bad Outcome?

George M. Dearborne

Staff Science Writer

San Fransisco—If you place a blackjack bet according to favorable odds but you lose, you made a good decision but experienced a bad outcome. On Tuesday, February 13, 1979, California Governor Williamson warned the San Francisco Bay Area of an imminent earthquake having the destructive force of the great 1906 earthquake. An earthquake did occur on February 15, well within the warning period of four days. The earthquake was felt by San Franciscans as a gently rolling motion that lasted for five seconds, but the damage was centered in the old mission town of San Juan Bautista. The partially restored mission—a favorite tourist attraction—was a total loss, as were several small commercial buildings. Sturctures were damaged as far north as San Jose and as far south as Salinas. However, an examination into the events of the three days in February disclosed that the official actions and private reactions of individuals put into motion by the warning resulted in more deaths and injuries than the earthquake itself.

Hearings were held in the State Office Building in San Francisco yesterday, April 24, 1979, by the California Legislative Special Joint Committee on Earthquake Prediction. These hearings probed into the events that preceded the warning and followed until well after the earthquake. Was Governor Williamson gambling, and like the card player, did he make a good bet that unfortunately resulted in a bad outcome? If so, did he have any choice in the matter? Governor Williamson did not testify at the hearings. The Special Joint Committee was established by the legislature on March 7. Yesterday's hearings were the first in a series of hearings and special studies to be conducted within the next year on the subject of earthquake prediction and the San Juan Bautista earthquake and warning. The magnitude 6.5 earthquake did very little to relieve the stresses that scientists feel are continually building up on the San Francisco Bay section of the great San Andreas fault. Similar stresses are building up in the Los Angeles area as well. The reaction to earthquake prediction expressed by many public and private officials at yesterday's hearings cast doubt on the usefulness of earthquake prediction information for public and private action in future earthquakes until the reliability is greatly improved.

Governor Williamson's 8 pm prime-time appearance on all of the major Bay Area television stations on February 13 warning of a major earthquake in the Bay Area within the next four days set off a chain of events that are by no means to be put to rest by the Joint Committee's hearings. The courts have a heavy backlog of lawsuits to settle, and the issue is sure to be hotly debated in next year's gubernatorial elections.

According to Douglas S. Buchannan, who is Chairman of San Mateo County's Board of Supervisors, Governor Williamson did not issue an official warnings in the legal sense that would bind local officials to take action. Supervisor Buchannan testified that "···these facts put the County in the position of having to act on the Governor's announcement, but having no state authority for doing so, and therefore of having to assume the full legal responsibility for the situation."

Mr. Buchannan related how San Mateo County engineers consulted with USGS scientists in Menlo Park and second-guessed the Governor on the probability of a great earthquake occurring. As a result, the County implemented its emergency plans on a selective basis.

Other public officials read the mandate of Governor Williamson's warning differently. Frederick L. Monitor, Chairman of Santa Clara County's Board of Supervisors, said that they "had no choice but to take the Governor's warning at face value. As a result we mobilized our full range of emergency services and implemented an emergency response plan appropriate to the situation, which we considered very serious."

Sara J. Buell, President of the San Francisco Board of Supervisors, agreed. "Because of the City Charter Amendments of 1976, the Mayor could not declare an emergency without approval of the Board of Supervisors. In an emergency session called shortly after the Governor's television appearance, the majority of the Supervisors present agreed the declaration of an emergency was the only course of action, although we had many reservations concerning liability, the availability of state and federal assistance, and so on."

San Francisco Mayor Nicholas Recuperio added that: "The basic emergency action with this type of short-term warning is to keep people out of potentially dangerous buildings. Therefore, except for certain vital functions, we ordered a shutdown of the central business district and evacuation of certain types of structures. In our announcement, which went out on television and radio, we made it clear that the one- and two-story wood-frame dwelling was the safest structure to be in during an earthquake. We also gave instructions about actions to be taken during and after an earthquake, such as finding shelter under a heavy table, or in a doorway, shutting off utilities—especially gas and electricity—and putting out small fires before they could grow. The following morning's newspapers also carried detailed instructions, but that was the last paper issued, because they were shut down for the duration of the emergency."

The USGS, whose earthquake prediction studies are directed from Menlo Park, gave testimony that was intended to emphasize that they presented the scientific facts to Governor Williamson uncolored about whether to issue a warning. Under questioning it did develop that, under the Freedom of Information Act, the USGS could not withhold the information and its policies, in line with executive orders and guidelines, would be to "take positive action to release this information to the public while it still could be acted on" if the Governor did not act. Governor Williamson was made aware of the USGS policies, they testified.

Dr. Robert G. Ainsworth, the Director of Earthquake Studies at USGS Headquarters, Reston, Virginia, responded to sharp questioning from Senator Ernest K. Norris, whose district includes parts of San Mateo County and San Francisco, by stating, "The simple facts are we felt confident enough that we could see the premonitors of an imminent earthquake on the few instruments that we had managed to install along a 10-kilometer section of the San Andreas fault that included San Juan Bautista. As you know, we have been fighting just to keep this program alive since 1975. The Parkfield earthquake last year did give us a shot in the arm, but not nearly big enough or soon enough. We did feel an obligation to point out to Governor Williamson the implications of what we didn't know; namely, that we didn't know what the rest of the so-called locked northern section of the fault was doing because we didn't have instruments. Even if we did have instruments that showed nothing, some of our people feel strongly that an earthquake of the size indicated anywhere along this part of the fault could trigger the release of accumulated stresses in the entire 'locked' section. In other words, we could put a lower bound on the size of the impending earthquake, but we couldn't specify the upper bound, except that we don't believe that the 1906 earthquake will ever be exceeded in this region."

The Chief of Earthquake Prediction Studies in Menlo Park, Dr. Henry T. Paine, elaborated on this point. "Our basis for predicting large earthquakes in this region is the retrospective prediction of the magnitude 7 Parkfield earthquake of September 4, 1978. This was a classic earthquake from the prediction point of view. We had dense enough instrumentation to see the earthquake premonitors along the entire length of the fault that ultimately broke. Furthermore, we obtained corroborating premonitors from three different kinds of instrumentation. However, that section of the fault has experienced creep and continual small earthquakes so it isn't considered 'locked.'

"Even though a magnitude 7.0 earthquake can be pretty damaging, the corresponding length of fault breakage is from one to two orders of magnitude, or 10 to 100 times, less than a 1906 type magnitude 8.3 earthquake. In other words, if a magnitude 7.0 earthquake represented 40 kilometers of fault breakage in a region, a magnitude 8.3 earthquake could represent as much as 400 kilometers."

Dr. Augstus S. Weiland, the USGS Chief Earthquake Seismologist in Menlo Park, made a telling point for their case: "What is not known in science can be as important as what is known. We knew that the state has an earthquake prediction evaluation committee, and so we told them what we knew as well as what we didn't know. They certainly have the expertise to sort it out."

Dr. Davis M. Keefer, Chairman of the state's Earthquake Prediction Evaluation Committee and an eminent seismologist from

State University, was reticent about discussing the deliberations of the Committee. He appeared under subpoena and made no formal statement. The minutes of the Committee's meeting, which were also subpoenaed, were made part of the record and are more revealing than were made part of the record and are more revealing than were Dr. Keefer's tangential answers to direct questioning. The meeting was held with four members present and with five members participating via a telephone conferencing hookup. The Committee agreed seven to two on the minimum prediction, but were unanimously against validating the speculation—as they characterized it—by the USGS about a large magnitude earthquake. Telephone company records showed that the entire conference lasted 17 minutes.

The prediction and its evaluation next went to the Governor's Seismic Advisory Committee at about 4:30 pm. The Chairman of that committee, Mayor Eric T. Fortune of Stockton, defined the function of the Seismic Advisory Committee: "To advise the Governor on the political and public-policy ramifications of a given prediction. As we saw the problem, we had little choice on the minimum earthquake," he added. "The question that we had to wrestle with was the possible maximum earthquake. We knew that the science panel had considered it of little validity as a scientific matter, but as a public matter we weren't too sure. And we were only given two hours to decide. Well, under the press of a 6:00 pm deadline, we reached a compromise position that an 'information only' bulletin be issued on the possible magnitude 8.3 earthquake."

Assemblyman Philander C. Meade, whose district includes Pasadena and Glendale in southern California, commented that "this was like Pilate washing his hands and leaving the decision to the mob." Mayor Fortune replied somewhat testily, "Not at all. We could have advised the Governor to set the scientific committee or the USGS up as the goat if he were wrong. We didn't feel that the information was firm enough to ask the 'feds' for help or to mobilize state forces, but it would have been criminal not to pass on the information, uncertain as it was. Sure, it put the decision to act or not on the locals. Some of them overreacted, some of them played it about right as it happened. Maybe it was luck. Maybe it was a matter of being smart."

Senator Leo Castelli of Santa Clara County, Chairman of the Joint Committee, took the time to sum up for the record his understanding of Governor Williamson's warning statement. "Those areas that could be affected by the magnitude 6.5 earthquake centered in San Juan Bautista in San Benito County were obliged to respond by implementing their emergency plans, while those areas that could be affected by a magnitude 8.3 earthquake, which is the entire Bay Area, could decide to implement emergency plans or parts thereof or not."

Several witnesses testified to the effect that the discretion exercised by local officials had on their individual lives and interests. The executive director of the Greater San Francisco Chamber of Commerce, Abram S. George, testified on the impact of closing down the central business district. "Using hindsight, of course, we can see that this was a big mistake. At the time, Mayor Recuperio painted a pretty dismal picture, and quite frankly, we were taken in. We agreed voluntarily to close down. It was apparent that the mayor had the backing of the Board of Supervisors, and we didn't want people facing the police or National Guard in order to get to work. I don't know what we would do the next time. When the cost of failures in this earthquake-prediction business exceeds the benefits of success, it seems to us that the exercise isn't worth the trouble. I think that we need pretty nearly 100 percent reliability in order to respond responsibly! Who knows how much the city lost during the three days it was shut down? Why, the gross city product is millions of dollars a day. Much of this loss was absorbed by businesses, much of it was passed on to employees in lost wages. Lost wages, by the way, that can't be spent for goods and services. I personally know many businessmen who have or are planning to bring suit for damages."

In his testimony Dr. Allen C. Wheaton, Jr., the noted economist at State University, stated that many of the business losses are merely transfers to other regions or most likely will be made up later by businesses in the Bay Area. He pointed out that many businesses increased as people took actions to mitigate the effects of the anticipated earthquake. "To the extent that these are not permanent improvements in property and so forth, they represent losses to individuals." But Professor Wheaton put this rhetorical question in the record. "How many failures can you afford if one success is instrumental in saving thousands of lives?" He likened the economic impact to an unplanned three-day weekend.

Mrs. Estelle Ferrell and two other representatives of the San Francisco Neighborhood Tenants Association described how they and their families were awakened in the middle of the night by policemen and ordered to move to Golden Gate Park. There was no transportation until morning, when Muni buses were pressed into service. When they arrived at Golden Gate Park, there was no shelter, and since it was raining, they refused to get off the buses. They ended up in a school in the Sunset District.

Mrs. Rosa Martinez of the tenant group testified, "My home withstood the 1906 earthquake, so as far as I'm concerned it has passed the test. Where are the poor and minorities expected to live? My home is all I can afford, but it's comfortable and surely looks more solid than those $100,000 cracker boxes that they're building today. They tear down all these fine old homes that we can afford and put up big apartments that we can't afford."

Samuel G. Blaine, President of San Francisco Senior Citizens Association, told of the plight of the elderly who live in old hotels and apartment buildings in or near the central business district. It's the only place they can afford on their income, which is often limited to Social Security payments. Also, they like to be near the facilities that the central city offers. "These people, who are often physically slow or handicapped and mentally easily confused, were put into a state of shock when they were ordered out of their homes," he said. "I'm not against such evacuation when necessary, but I am against the way it was carried out, and I am fundamentally against the social conditions that put these people in substandard buildings in the first place."

Mr. and Mrs. Robert W. Redman III appeared in sharp contrast to the tenant group and contingent of elderly attending the hearings. They live in the plush Bayview Towers and testified that they had decided to evacuate their quarters when they heard an eminent local structural engineer comment, on an "earthquake special" on television following Governor Williamson's announcement, that as many as twenty of San Francisco's newer high-rise buildings could collapse in a repeat of the 1906 earthquake. They decided to head east toward the safety of their condominium apartment at Lake Tahoe. But they found that they only had a quarter tank of gas. Since there were long lines at all of the stations in their neighborhood, they thought that they would have more luck in Oakland. They got stuck in a traffic jam on the Bay Bridge and ran out of gas in East Oakland. Discouraged because the stations in Oakland were out of gas, they returned to San Francisco on a BART train. Describing their experience, Mr. Redman said, "It was scary as hell sitting on the Bay Bridge knowing that at any time it could start swaying and dump you into the inky black Bay."

Chester L. Joiner, Executive Director of the Bay Bridge Authority, related how, with the help of the Highway Patrol, traffic was redirected on the upper half of the Bay Bridge from incoming to outgoing. This eased the traffic jam for a time until the freeways in Oakland backed up.

Mayor S. David Hall of Oakland, who opted not to warn his community of a possible magnitude 8.3 earthquake, described for the joint committee how his city had become the unintended "host" for thousands of West Bay residents. As a result of accidents, abandoned and stalled vehicles, and exhaustion of gasoline supplies, many would-be evacuees got no farther than Oakland and its suburbs in the East Bay. All public buildings, schools, and churches were opened, and many private citizens offered space in their homes. Mayor Hall admitted that many Oakland residents joined the stream of evacuees as a result of the television broadcasts of the Governor and the Mayor of San Francisco and from seeing the streams of evacuees from the West Bay.

Captain Stanley Hathaway of the California Highway Patrol estimated that 49 deaths could be attributed to the emergency evacuation. Twenty-seven were killed in traffic accidents between vehicles, six pedestrians were run down, seven died from carbon monoxide poisoning; there were also five apparent suicides and four homicides.

Dr. Miles E. Lessenco, the San Francisco Director of Public Health, said that the emergency was too short for a serious public health problem to develop. Because of the inclement weather, seven persons, most of whom were elderly or ill, died of exposure attempting to camp out on the road, in parks, or in their backyards. There were 37 heart attacks during the first 12 hours after the warning. For a time traffic jams were so bad that ambulance service was virtually at a standstill. Many of the persons who were attempting to evacuate went as long as 36 hours without a meal.

San Benito County, where the earthquake occurred, was also plagued with evacuees. The situation got so bad, according to Michael R. Suarez, Chairman of the Board of Supervisors, that they feared that they would not be able to render assistance to victims. Supervisor Suarez testified that, "After consultation with the Board of Supervisors, the Sheriff established checkpoints in the northern and western parts of the county where the heaviest traffic from Route 101 was. The deputies attempted to discourage every one who was not a resident of the county from entering. The voluntary approach worked very well in reducing the numbers of people entering the County to a trickle during the night of Wednesday, February 14."

Supervisor Suarez further testified that"···the Sheriff estimated that by Friday morning, there were 10,000 additional people in the County. This is half again our total County population, which is just under 20,000." Suarez estimated that the earthquake damage in the County was $ 30 million. The majority of this was in the water system and agricultural lands. He put the toll at 21 persons injured, and he guessed that the actions taken as a result of the warning had saved "between ten and twenty lives."

The Superintendent of San Benito County Schools, Harold G. Leslie, testified that the schools, which were being occupied by evacuees from the north at the time, performed very well. All of the county's schools are Field Act schools, which means that they are built to resist earthquake forces.

Although there was confusion over the proper course of action within government circles, this was not the case at the giant Pacific Gas and Electric Company, according to Frederick T. Winslow, Vice President, Gas Operations. "We have been following the science of earthquake prediction for some time," stated Mr. Winslow. "We learned quite a bit at a very high price about geology and seismology during our attempts to site nuclear power plants in California. We and our consultants just didn't put any credibility in the prediction of a 1906-type earthquake. Even if we did, are we to deprive homes, hospitals, and other vital facilities of gas for heat and cooking, especially during winter? Our crews were alerted and standing by, and we could have cut off all primary gas mains within a few minutes after the earthquake. If the homeowners and other gas users would cooperate and close their service valve at the same time, we would have no problems with gas-caused fires. We run a greater risk from unauthorized emergency forces shutting off our high-pressure mains."

Assemblyman Carrasco asked John H. Torkeley, Regional Manager of the Golden State Insurance Company, why a homeowner should be concerned with fire if his home is damaged by an earthquake. "He generally has fire insurance, but not earthquake insurance," stated Carrasco. "Didn't the payment of fire insurance claims after 1906 largely rebuild San Francisco?" Torkeley replied that people have a deep emotional involvement in their homes. "People put a lot of work in their home; their memories are tied up in it; their valuables and treasured possessions are in it; so we don't find people torching their homes even in situations such as you describe," said Torkeley. "Businesses, especially small and marginally profitable ones, are a different story," Torkeley was quick to add.

When asked whether earthquake insurance was still available, Torkeley replied, "Indeed it is. You know the Bay Area still faces a major earthquake; nothing has really changed. But my door isn't being beaten down by potential customers."

On the question of the effect of reliable earthquake prediction on the availability of earthquake insurance, Torkeley stated, "Reliable and accurate long-term predictions would force us out of that business, but we aren't there yet; it's still uncertain and so a risk for the property owner. Insurance is designed to help the individual with the unexpected."

"The warning affected us in two ways," stated Arnhold T. Friendly, Vice President of Banking Service for the Bank of America. "We had a moderate run on our banks for cash on the day after the warning. It's not that people didn't have faith in the banks, but they were concerned that if an earthquake did occur there would be a strictly cash economy. That brings me to the next point," Friendly added. "Banking today operates on data and the ability to process it. Our data is fairly secure, we believe, but we don't know whether our data processing machines would remain functional after an earthquake; or for that matter whether we would have employees available to operate them. Therefore, we started transferring some of our data to Los Angeles for processing immediately after the warning. Some of it was physically transferred by aircraft. Some was transferred by telephone hookup. We felt we were on top of the situation at all times."

Standard Oil's large refinery at Richmond in the East Bay kept humming all during the situation. Charles J. Veitch, the General Manager of the Richmond Refinery, said, "It takes five days to properly shut down a large refinery complex. We have emergency shutdown procedures that do some damage to some refinery components, but we felt that we could wait until and if the earthquake struck to see what damage would occur. The processing equipment is pretty rugged. We may have problems with large storage tanks, but there wasn't anything we could do about them."

Safeway Stores' Regional Manager Joseph K. Machado told the committee about the difficulty of keeping some foods on the shelves. "People bought especially large amounts of canned goods and packaged dry goods. People normally do their shopping during the latter part of the week, but sales after the earthquake warning were unusually heavy. Many stores were virtually cleaned out of their stocks of these items. However, they were all restocked within a week."

At no time during the emergency was the National Guard called out. Major General Clyde Anderson Stevenson testified: "We were standing by awaiting orders from Governor Williamson, but they never came. If the large earthquake had struck, we would have been needed."

California Office of Emergency Services Director William P. Lampson echoed General Stevenson's remark. "We are set up to coordinate and direct the postdisaster efforts of county and city units of government. We were ready and waiting, but fortunately we weren't needed."

The next hearings of the Joint Committee will be held at 9:00

am on May 8 in the Resources Building Auditorium in Sacramento.

12.5.4   Cancellation of earthquake warning

What would happen when an earthquake warning was issued and no earthquake occurred during a period which is supposed to be dangerous? The warning must be either cancelled or postponed although no guidelines about how to make such a judgment has as yet been established.

According to the Large-scale Earthquake Countermeasures Act, which will be outlined in Section 13.1, the Prime Minister shall, upon consultation with the cabinet, cancel the warnings statement when he receives additional prediction information from JMA's Director General and he recognizes that the possibility of the occurrence of the earthquake has disappeared. It is not known, however, how such information can be provided. It is of course the responsibility of the Prediction Council to put forward the information. But the Council has no manual for such a purpose.

Should the anomalies, which provide the basis of earthquake warnings statement, somehow disappear, it might be possible to cancel the warning. It appears to the author, however, that a more objective way for cancelling would be provided by taking the probability of earthquake occurrence into account. Such a probability may be estimated on the basis of earthquake precursors of the second kind. As the author argued in relation to the time window for dangerous periods of earthquake occurrence in Section 9.2, it may be possible to assume that the signals have been false when the probability exceeds a prescribed high value.

In an actual case, however, it would be terribly difficult to estimate the said probability very accurately. It cannot be helped, therefore, to make a synthesized judgment by taking various factors into account at the present stage of earthquake prediction technology. In short, cancellation of prediction information is as difficult as issuance of the information.

12.6   Recent Contributions to the Problem of Public Reaction Against an Earthquake Warning

The subject treated in this chapter tended to draw the keen

attention of sociologists quite recently. The 1979 UNESCO Symposium on Earthquake Prediction (see Section 1.5) has no doubt stimulated interested sociologists.

Although a few new reports (TURNER et al., 1979; HUTTON et al., 1981; INSTITUTE OF JOURNALISM, 1979a, b) reached the author, he regrets that no account of them can be cited here because preparation of the manuscript of this book had been completed at that time.

CHAPTER 13

LEGISLATION RELEVANT TO EARTHQUAKE PREDICTION AND WARNING

13.1 Large-Scale Earthquake Countermeasures Act

Mention has often been made in the book of the Large-scale Earthquake Countermeasures Act, which was enacted in December, 1978 in Japan. Subsection 10.2.1(5) states in fair detail how the legislation was formulated. The author will outline briefly the contents of the act in this section (RIKITAKE, 1979c). The full text of the act is given in Appendix I.

The Large-scale Earthquake Countermeasures Act aims at mitigating earthquake hazards by providing the measures to be taken by the national government, local governments, and public and private corporations in case a large earthquake is predicted. In the course of deliberation of the act at the Diet, it was made clear that a "large-scale earthquake" was to be considered by the law for the time being as an earthquake having a magnitude around 8. If the techniques of earthquake prediction are developed to a higher level in the future, the definition could be extended to quakes of magnitude 7 or thereabout.

According to the law, an area will be designated by the Prime Minister, who consults the Central Disaster Prevention Council and hears the views of prefectural governors, to be an "area under intensified measures against earthquake disaster" when a long-term prediction that a large earthquake will hit there in the near future is made and when considerable damage due to the quake is expected.

CCEP pointed out, in 1976, that the possibility of having an earthquake of magnitude 8 or so in the Tokai area exists. Accordingly practical legislation was enacted to cope with the feared Tokai earthquake. A special committee of experts attached to the Central Disaster Prevention Council worked on forming a source model of the expected earthquake and estimating the intensities of ground vibration expected from the model.

The committee put forward the most likely source model which is shown in Fig. 13.1 in which the horizontal projection of a fault plane that would slip at the time of occurrence of the expected earthquake is indicated. It is assumed that the plane dips approximately to the west with an angle amounting to 20–30 degrees from the horizontal plane. The earth crust to the west of the Suruga trough would thrust up a

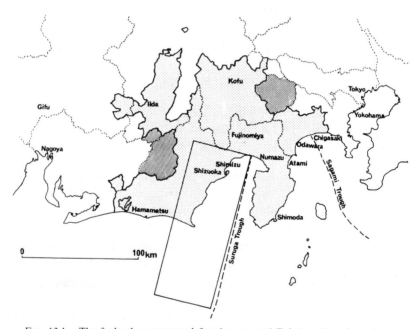

FIG. 13.1. The fault plane supposed for the expected Tokai earthquake and the "area under intensified measures against earthquake disaster" as designated by the Prime Minister on Aug. 7, 1979. The shaded area has been originally proposed by the Central Disaster Prevention Council, while the hatched areas are added after consulting with prefectural governors.

few meters. It is further assumed that the rupture would first occur at the southernmost part of the plane and would be propagated to the north covering all the fault plane within a few tens of seconds.

On the basis of such a source model, some of the members of the committee, who are working on earthquake engineering, estimated the expected intensities of ground motion taking the effect of local conditions of ground into account. The area in which the intensity would exceed VI in the JMA scale was thus obtained. It was first proposed to designate the area consiting of cities, towns and villages, which are covered totally or partially by the area of intensity VI, to the "area under intensified meaures against earthquake disaster". As a result the area became the shaded one in Fig. 13.1.

Upon consultation with prefectural governors, some modification was made by adding a number of municipalities to the original plan. For instance, the Governor of the Shizuoka Prefecture wanted to include the remaining three villages in the designated area because it is easier to take measures for disaster prevention by handling all the prefecture as a whole. The hatched areas in Fig. 13.1 are those added in this way.

The actual designation of the area was made on August 7, 1979. As no account of landslides related to sloping topography, effect of long-period motion and quicksand effect, was taken this time, it is possible to extend later the designation to some areas by taking the above effects into account.

The State shall intensify seismological and other observations over the "area under intensified measures against earthquake disaster" thus designated in order to detect possible precursory effects for short-term and imminent prediction.

The Central Disaster Prevention Council, chaired by the Prime Minister, shall, of course, formulate a basic plan of disaster prevention. Local governments, public bodies, and the like in the designated area, are also obliged to formulate intensified plans of earthquake disaster prevention, such as providing open spaces, evacuation routes, and including training, education, and exercise for the public.

Organizations in the area, such as hospitals, theaters, department stores, hotels, petroleum refineries, high-pressure gas and explosives factories, railway and bus companies, and the like are asked to formulate a "short-term plan of earthquake disaster prevention" within 6 months from the date of designation. National and local governments will naturally have to take extra budgetary action for earthquake hazard reduction of public systems.

Should short-term (probably ranging from several hours to a few days) prediction information be conveyed to the Prime Minister from the Prediction Council via the Director General of JMA, the Prime Minister, in consultation with Cabinet members, would issue an "earthquake warnings statement". In that case a "national head-quarters for earthquake disaster prevention" is immediately set up, and the Prime Minister is appointed as the general director.

The general director is empowered to take emergency actions which cannot be taken at ordinary times. For instance, he could ask the Self-Defense Force to come to the earthquake-threatened area and take necessary action to keep the public in order. He could also ask local public bodies to take appropriate action necessary for disaster prevention.

Prefectural and municipal headquarters, directed by governors and mayors, are also to be set up. They are empowered to take necessary actions for traffic control, crime prevention, evacuation, and so on. It could, therefore, become possible to shut down oil refineries, nuclear power generation, and the like and to close highways or stop railways, including the Shinkansen, if necessary. If required, it is possible to take over privately owned materials for emergency use, with later compensation. Even some penal regulations for those who violate the law are included in the act.

When the outline of the act was presented at the April 1979 UNESCO Symposium on Earthquake Prediction, the act attracted much attention of the audience, which included many sociologists. The author should here like to emphasize the important point, that realistic legislation, which relies on earthquake prediction information, has been made in Japan for the first time in history.

Although the act entirely relies on earthquake prediction tech-nology, no practical measures for promoting earthquake prediction are included in the act except article 33 that reads "In order to predict the occurrence of earthquakes, the State shall make efforts to improve the facilities and equipments for seismological observations and surveys. At the same time, the State shall endevor to improve the system of research, promote research and diffuse its results for the advancement of science and technology which will contribute to earthqake prediction."

It is doubtful, however, if such encouragement, which is only spritual, will be all that useful for a dramatic development of earthquake prediction. Much more concrete measures for promoting science and technology of earthquake prediction should be hurriedly taken in Japan. For instance, the setting up of a powerful organ for earthquake prediction as discussed in Subsection 6.1.7 should be seriously considered in order to make the Large-scale Earthquake Countermeasures Act work efficiently.

13.2   Earthquake Hazards Reduction Act of 1977

The "Earthquake Hazards Reduction Act of 1977" was enacted in the U.S.A. on October 7, 1977. Appendix II is the full text of the act.

It is intended in the act to clarify the general measures for reducing earthquake hazards, so that, while importance of earthquake prediction is, of course, stressed in the act, the measures such as those closely related to earthquake engineering, which are directly concerned with mitigation of earthquake disaster, occupy the main part of the law.

According to the law, the President shall put forward a program for mitigating earthquake disaster. Within 30 days after the date of enactment, the President shall designate the Federal department, agency, or entity reaponsible for the development of the implementation plan for prediction and warning of earthquake. The President shall submit to the appropriate authorizing committees of the Congress the implementation plan within 210 days after the date of enactment.

Within 300 days after the date of such enactment, the President also shall designate the Federal department, agency or inter-agency group which shall have primary responsibility for the development and implementation of the earthquake hazards reduction program, assign and specify the role and responsibility of each appropriate Federal department, agency, and entity with respect to each object and element of the program, establish goals, priorities and target dates for implementation of the program, provide a method for cooperation and coordination with, and assistance (to the extent of available resources) to, interested governmental entities in all States, particularly those containing areas of high or moderate seismic risk, and provide for qualified staffing for the program and its components.

The President shall submit an annual report to the appropriate authorizing committees in the Congress describing the status of the program, and describing and evaluating progress achieved during the preceding fiscal year in reducing the risks of earthquake hazards within 90 days after the end of each fiscal year.

There are also authorized to be appropriated to the President, USGS and NSF funds for purposes of carrying out the responsibilities that may be assigned to them under the Act not to exceed a certain amount.

The Earthquake Hazards Reduction Act of 1977 seems to be enacted in order to intensify general measures for mitigating earthquake disaster, although the Act is not as far-reaching as the Japanese Large-scale Earthquake Countermeasures Act which aims at coping with a great earthquake that would actually occur in the foreseeable future. The seismicity in the U.S.A. is lower than that in Japan. The frequency of great earthquakes is especially small in the U.S.A. In these circumstances it may be that the American public is not concerned about earthquake disaster so much as the Japanese. Even so, the point that legislation such as the present Act has already been formulated should be appreciated. The author, will watch with interest, developments of American measures for earthquake prediction and warning, for future progress on the basis of the Act.

LARGE-SCALE EARTHQUAKE COUNTER-MEASURES ACT

Purpose

Article 1. The purpose of this Act is to protect life and property of the citizens from the hazards of earthquakes through the intensification of earthquake disaster prevention measures by taking necessary steps regarding the designation of areas under intensified measures against earthquake disaster. establishment of a seismological observation system, and other matters relating to the establishment of an earthquake disaster prevention system, and also regarding short-term prevention measures against earthquake disaster and other related matters, thus contributing to the preservation of social order and the securing of public welfare.

Definitions

Article 2. Terms employed in this Act shall be defined as below:
1) *Earthquake disaster* means damage directly caused by an earthquake and any other damage caused by tsunami, fire, explosion and any other unusual phenomena which may follow the earthquake.
2) *Earthquake disaster prevention* means the planning to prevent an earthquake disaster or to reduce the damage which may be caused by the disaster if it should occur.
3) *Earthquake prediction information* means the information concerning earthquakes prescribed under Article 11-2, Paragraph 1 of the Meteorogical Service Law (Law No. 165, 1952) and the information concerning a new situation prescribed under Paragraph 2 of the same Article.
4) *Areas under intensified measures against earthquake disaster* means those

areas designated under the provision of Paragraph 1 of the next Article.

5) *Designated administrative organs* means those administrative organs designated under the provision of Article 2-(3) of the Disaster Countermeasures Basic Law (Law No. 223, 1961).

6) *Designated local administrative organs* means those local administrative organs designated under the provision of Article 2-(4) of the Disaster Countermeasures Basic Law.

7) *Designated public corporations* means those public corporations designated under the provision of Article 2-(5) of the Disaster Countermeasures Basic Law.

8) *Designated local public corporations* means those local public corporations designated under the provision of Article 2-(6) of the Disaster Countermeasures Basic Law.

9) Earthquake disaster prevention plans means a basic plan of earthquake disaster prevention, an intensified plan of earthquake disaster prevention and a short-term plan of earthquake disaster prevention.

10) *Basic plan of earthquake disaster prevention* means a basic plan to be prepared by the Central Disaster Prevention Council for the area under intensified measures against earthquake disaster.

11) *Intensified plan of earthquake disaster prevention* means those sections of the following plans prescribed for matters under Article 6, Paragraph 1: the operational disaster prevention plan prescribed under Article 2-(9) of the Disaster Countermeasures Basic Law, the local disaster prevention plan prescribed under Article 2-(10) of the same Law and the disaster prevention plan for petroleum complex and others prescribed under Article 31, Paragraph 1 of the Disaster Prevention Law for Petroleum Complex and Others (Law No. 84, 1975).

12) *Short-term plan of earthquake disaster prevention* means a plan regarding short-term prevention measures against earthquake disaster to be prepared by those designated under the provision of Article 7, Paragraph 1 or Paragraph 2 of this Act.

13) *Earthquake warnings statement* means the warnings statement regarding earthquake disaster issued by the Prime Minister under the provision of Article 9, Paragraph 1.

14) *Short-term prevention measures against earthquake disaster* means those short-term measures to be taken for the prevention of the large-scale earthquake disaster during the period from the time of the issuance of a warnings statement until the occurrence of the earthquake for which the statement has been issued or until such a possibility has disappeared.

Designation of Areas under Intensified Measures against Earthquake Disaster, etc.

Article 3. Because serious disaster may ensue when a large-scale earthquake occurs within the earth crust where a large-scale earthquake is particularly hazardous, the Prime Minister shall designate the area where it is necessary to take intensified measures for earthquake disaster prevention as an area under intensified measures against earthquake disaster (hereafter referred to as "intensified area").

2. The Prime Minister shall consult the Central Disaster Prevention Council prior to making such a designation as prescribed under the provision of the above Paragraph.

3. The Prime Minister shall hear the views of the prefectural governors concerned, prior to making such a designation as prescribed under the provision of Paragraph 1, who are to hear the views of the mayors of municipalities (cities, towns and villages) concerned, prior to presenting their views to the Prime Minister.

4. The Prime Minister, upon making such designations as prescribed under the provision of Paragraph 1, shall make public announcements accordingly.

5. The provisions under the preceding three Paragraphs are applied correspondingly to the case of the Prime Minister cancelling such designations of intensified areas as prescribed under the provision of Paragraph 1.

Intensification of Seismological Observations and Surveys Regarding Intensified Areas

Article 4. In order to predict the occurrence of a large-scale earthquake related to the intensified areas and to prevent or reduce earthquake disaster, the State shall systematically intensify seismological observations and surveys, carrying out incessant observations of geophysical and hydrological phenomena and intensifying geodetic surveys of land and sea bottom (referred to as "surveys" in Article 33).

Basic Plan of Earthquake Disaster Prevention

Article 5. When an intensified area is designated under the provision of Article 3, Paragraph 1, the Central Disaster Prevention Council shall formulate a basic plan of earthquake disaster prevention for the intensified area and promote its implementation.

2. The basic plan of earthquake disaster prevention is to formulate a basic policy of the State regarding earthquake disaster prevention following the issuance of earthquake warnings statement, the basis for an intensified plan of earthquake disaster prevention and a short-term plan of earthquake disaster prevention and other matters prescribed under the cabinet order of the Act.

3. The provision of Article 34, Paragraph 2, of the Disaster Countermeasures Basic Law is applied correspondingly to the case of formulating or amending the basic plan of earthquake disaster prevention prescribed under Paragraph 1.

Intensified Plan of Earthquake Disaster Prevention

Article 6. With the designation of an intensified area under the provision of Article

3, Paragraph 1, chief officers of designated administrative organs prescribed under the provision of Article 2-(9) of the Disaster Countermeasures Basic Law (or chief officers of local administrative organs, within the matters entrusted by the chief officers of the designated administrative organs) and designated public corporations (or local public corporations, within the matters entrusted by the designated public corporations) shall, in the operational disaster prevention plan prescribed under Article 2-(9) of the Disaster Countermeasures Basic Law, make arrangements itemized below. The same applies to the local disaster prevention council and those councils prescribed under the provision of Article 21 (or the mayors of the municipalities concerned in the absence of the council) for the local disaster prevention plan prescribed under the provision of Article 2-(10), the headquarters for disaster prevention for petroleum complex and others prescribed under the provision of Article 27, Paragraph 1 of the Disaster Prevention Law for Petroleum Complex and Others (referred to as the headquarters for disaster prevention for petroleum complex and others under Article 28, Paragraph 2) and the council of the disaster prevention headquarters prescribed under the provision of Article 30, Paragraph 1 for the disaster prevention plan for petroleum complex and others under the provision of Article 31, Paragraph 1.

1) Matters relating to short-term prevention measures against earthquake disaster.

2) Matters relating to those facilities prescribed under the cabinet order such as places of refuge, evacuation routes and fire-fighting facilities to be provided promptly for the prevention of the large-scale earthquake disaster.

3) Matters prescribed under the cabinet order relating to disaster prevention exercises and other disaster prevention measures against the large-scale earthquake.

2. The intensified plan of earthquake disaster prevention shall be based on the basic plan of earthquake disaster prevention.

Short-Term Plan of Earthquake Disaster Prevention

Article 7. Those who shall administer or operate the facilities or enterprises within the intensified area, which are listed below and prescribed under the cabinet order (excluding those prescribed under Paragraph 1 of the preceding Article) shall formulate a short-term plan of earthquake disaster prevention for each facilities or enterprises in advance.

1) Hospitals, theaters, department stores, hotels and other facilities used by an unrestricted and large number of people.

2) Those facilities which manufacture, store, process or handle petroleum, explosives, high-pressure gas and other materials prescribed under the cabinet order.

3) Private railways and other enterprises relating to passenger service.

4) In addition to those listed above, other important facilities or enterprises for which earthquake disaster prevention measures are considered to be necessary.

2. Those, who administer or operate the facilities or enterprises in the preceding cabinet order within the area upon the designation of an intensified area under the provision of Article 3, Paragraph 1 (excluding those prescribed under the provision of Paragraph 1 of the preceding Article), shall formulate a short-term plan of earthquake disaster prevention within six months from the date of the designation.

3. If it becomes necessary to amend the short-term plan of earthquake disaster prevention because of the expansion of the facilities or changes in the operations of the enterprises, those who formulated the plan shall alter the plan without delay.

4. The short-term plan of earthquake disaster prevention is to make arrangements regarding short-term prevention measures against earthquake disaster for the facilities or enterprises concerned and other matters prescribed under the cabinet order.

5. The short-term plan of earthquake disaster prevention shall not contradict or interfere with the intensified plan of earthquake disaster prevention.

6. Those prescribed under the provision of Paragraph 1 or 2 who formulated the short-term plan of earthquake disaster prevention are required to report it to the prefectural governors concerned without delay and send the copy to the mayors of municipalities concerned. The same applies when the plan has been amended.

7. If those prescribed under the provision of Paragraph 1 or 2 fail to comply with the above requirement to report, the prefectural governors concerned may advise them to do so within a specified period of time.

8. If those who have received the advice prescribed above fail to report within the specified period of time, the prefectural governors may make the fact public.

Special Cases in Formulating the Short-Term Plan of Earthquake Disaster Prevention

Article 8. If those prescribed under the provision of Paragraph 1 or 2 of the preceding Article made arrangements based on legal provisions in the plans and the regulations listed below, regarding the matters prescribed under Paragraph 4 of the above Article for facilities or enterprises prescribed under Paragraph 1 of the same Article, those arrangements made regarding the matters concerned (referred to as "regulations of earthquake disaster prevention") are to be regarded as the short-term plan of earthquake disaster prevention and this Act shall be applied to these arrangements.

1) The fire-fighting plan prescribed under the provision of Article 8, Paragraph 1 or Article 8-2, Paragraph 1 of the Fire Service Law (Law No. 186, 1948) or the regulations of fire prevention prescribed under Article 14-2, Paragraph 1 of the same Law.

2) The regulations of disaster prevention prescribed under Article 28, Paragraph 1 of the Explosives Control Law (Law No. 149, 1950).

3) The regulations of disaster prevention prescribed under Article 26, Paragraph 1 of the High Pressure Gas Control Law (Law No. 204, 1951).

4) The safety regulations prescribed under Article 30, Paragraph 1 (including the corresponding application under Article 37-7, Paragraph 3) of the Gas Utility Industry Law (Law No. 51, 1954).

5) The safety regulations prescribed under Article 52, Paragraph 1 of the Electricity Utility Industry Law (Law No. 170, 1964).

6) The safety regulations prescribed under Article 27, Paragraph 1 of the Petroleum Pipeline Business Law (Law No. 105, 1972).

7) The regulations of disaster prevention prescribed under Article 18, Paragraph 1 of the Disaster Prevention Law for Petroleum Complex and Others.

8) Those prescribed under the order of the Prime Minister's Office as corresponding to the above plans and regulations.

2. Those who formulated the regulations of earthquake disaster prevention are required, irrespective of the provision of Paragraph 6 of the preceding Article, to send the copy to the mayors of the municipalities concerned by the cabinet order. The same applies when the regulations of earthquake disaster prevention have been amended.

Earthquake Warnings Statement

Article 9. When the Prime Minister, upon receipt of earthquake prediction information from the Director-General of the Meteorological Agency, recognizes that the implementation of short-term measures against earthquake disaster is in urgent need, he shall, upon consultation with the Cabinet, issue an earthquake warnings statement. At the same time, the Prime Minister shall take the following steps.

1) To notify publicly the residents, sojourners and others, and public and private bodies within the intensified area (hereafter referred to as "residents") that they shall be prepared for the emergency.

2) To inform the designated public corporations and the prefectural governors concerned that the short-term prevention measures against earthquake disaster are to be taken in accordance with laws or the intensified plan of earthquake disaster prevention.

2. Upon issuing an earthquake warnings statement, the Prime Minister immediately shall take such steps as to inform the citizens of the contents of the earthquake prediction information. In this case the Prime Minister shall make the Director-General of the Meteorological Agency explain technical matters regarding the earthquake prediction information.

3. When the Prime Minister recognizes that the possibility of the occurrence of the earthquake which has been warned has disappeared upon receipt of additional earthquake prediction information from the Director-General of the Meteorological Agency after issuing the earthquake warnings statement, he shall, upon consultation

with the Cabinet, cancel the warnings statement. At the same time, the Prime Minister shall notify publicly those prescribed under Paragraph 1-(1) accordingly and inform those prescribed under Paragraph 1-(2) that the implementation of the prescribed measures is to be terminated.

Establishment of the National Headquarters for Earthquake Disaster Prevention

Article 10. Upon issuing an earthquake warnings statement, the Prime Minister shall establish temporarily, within the Prime Minister's Office, the National Headquarters for Earthquake Disaster Prevention (hereafter referred to as the "Headquarters") irrespective of the provision of Article 8 of the National Government Organization Law (Law No. 120, 1948).
2. The name of the Headquarters, its jurisdiction, location and duration shall be decided by the Prime Minister upon consultation with the Cabinet.

Organization of the Headquarters

Article 11. The national headquarters for earthquake disaster prevention shall be headed by a general director (hereafter referred to as "the General Director" to Article 13), who shall be the Prime Minister (in case the Prime Minister is not able to carry out his duties, a Minister designated in advance by the Prime Minister).
2. The General Director shall supervise the affairs and members of the staff of the Headquarters.
3. The Headquarters shall have a Vice General Director, staff members and other officials.
4. The Vice General Director shall be a Minister.
5. The Vice General Director shall assist the General Director and act for the latter in case the latter is not able to carry out his duties.
6. The staff members and other officials of the Headquarters shall be appointed by the Prime Minister from officials of designated administrative organs or chief officers or officials of designated local administrative organs.

Task of the Headquarters

Article 12. The Headquarters shall perform the tasks listed below:
1) To integrate and coordinate those short-term prevention measures against earthquake disaster or those emergency measures against earthquake disaster prescribed under the provision of Article 50, Paragraph 1 of the Disaster Countermeasures Basic Law (hereafter referred to as "the short-term prevention measures against earthquake disaster, etc.") taken within the jurisdiction by chief officers of designated administrative organs and designated local administrative organs, heads of local public bodies and other executive organs, designated public corporations and designated

local public corporations.

2) Matters within the power of the General Director prescribed under the provision of Article 28, Paragraph 1 of the Disaster Countermeasures Basic Law corresponding to the provision under the next Article and Article 15.

3) In addition to those listed above, those matters within the power of the Headquarters as prescribed under legal provisions.

Power of the General Director

Article 13. In order to carry out the short-term prevention measures against earthquake disaster, etc. appropriately and quickly, the General Director may, if necessary, give necessary instructions, to the extent of the necessity, to chief officers of designated local administrative organs, heads of local public bodies and other executive organs, designated public corporations and designated local public corporations concerned.

2. In order to carry out the short-term prevention measures against earthquake disaster appropriately and quickly, the General Director may, if necessary, request Minister of State for Defense to dispatch the Self-Defense Forces as prescribed under the provision of Article 8 of the Self-Defense Forces Law (Law No. 165, 1954) to aid in the implementation of the short-term prevention measures against earthquake disaster.

Abolition of the Headquarters

Article 14. The Headquarters shall be abolished with the establishment of the headquarters for major disaster control as prescribed under the provision of Article 24, Paragraph 1 of the Disaster Countermeasures Basic Law or the headquarters for emergency disaster control as prescribed under the provision of Article 107, Paragraph 1 of the same Law in respect of the earthquake disaster anticipated, or with the expiration of the period of establishment of the Headquarters.

Corresponding Application of the Disaster Countermeasures Basic Law to the Headquarters

Article 15. The provisions of Article 24, Paragraph 3, Article 27 and Article 28, Paragraph 1 of the Disaster Countermeasures Basic Law shall be applied correspondingly to the case of the establishment of the Headquarters. In this case, the "emergency measures against disaster" in Article 27, Paragraph 1 of the same Law shall be understood as "emergency measures against disaster or short-term prevention measures against earthquake disaster prescribed under Article 2-(14) of the Large-Scale Earthquake Countermeasures Act."

Establishment of the Prefectural Headquarters for Earthquake Disaster Prevention and of the Municipal Headquarters for Earthquake Disaster Prevention

Article 16. Upon the issuance of an earthquake warnings statement, prefectural governors of the intensified area shall establish the prefectural headquarters for earthquake disaster prevention (hereafter referred to as "the prefectural headquarters"), and mayors of municipalities shall establish the municipal headquarters for earthquake disaster prevention (hereafter referred to as "the municipal headquarters").

Organization and Task of the Prefectural Headquarters

Article 17. The prefectural headquarters shall be headed by a general director, who shall be the prefectural governor.

2. The prefectural headquarters shall have vice general directors, staff members and other officials.

3. The vice general directors of the prefectural headquarters shall be appointed by the prefectural governor from the staff members of the prefectural headquarters.

4. The vice general directors shall assist the general director and act for the latter in case the latter is not able to carry out his duties.

5. The staff members shall consist of those listed below:

1) The chief officers of designated local administrative organs exercising jurisdiction over the whole or part of the prefectural area, or its officials named by the chief officers.

2) The Commanding Generals of the Ground Self-Defence Force exercising jurisdiction over the prefectural area or the chief officers of units or organs named by the Commanding Generals.

3) The chief educator of the board of education of the prefecture.

4) The Superintendant-General of the Tokyo Metropolitan Police Department or the directors of the prefectural police headquarters concerned (referred to as directors of police headquarters" under Article 23, Paragraph 5).

5) Those prefectural officials named by the prefectural governors.

6) Those officials of municipalities and fire-fighting units within the prefectural areas appointed by the prefectural governors concerned.

7) Those officers or officials of designated public corporations or local public corporations operating in the prefectural areas concerned appointed by the prefectural governors concerned.

6. Those officials other than vice general directors and staff members of the prefectural headquarters shall be appointed from prefectural officials by the prefectural governors concerned.

7. The prefectural headquarters shall perform the tasks listed below:

1) To coordinate those short-term prevention measures against earthquake

disaster, etc. implemented by chief officers of designated local administrative organs, mayors of municipalities and other executive organs, designated public corporations and designated local public corporations in the prefectural area.

2) To implement short-term prevention measures against earthquake disaster, etc. in the prefectural areas concerned and to promote the implementation.

3) Matters within the power of the general director of the prefectural headquarters for earthquake disaster prevention as prescribed under the next Paragraph.

4) In addition to those listed above, other matters within its power as prescribed under various laws or cabinet orders based on them.

8. In order to carry out short-term prevention measures against earthquake disaster, etc. in the prefectural areas concerned, the general directors of the prefectural headquarters may, to the extent of the necessity, give instructions to the prefectural police or the prefectural board of education.

9. Other necessary matters than those listed above, regarding the prefectural headquarters, shall be prescribed under the prefectural bylaws.

10. While the prefectural headquarters are established, the prefectural disaster prevention councils prescribed under Article 14, Paragraph 1 of the Disaster Countermeasures Basic Law shall not, irrespective of the provision of Paragraph 2 of the same Article, perform its task regarding earthquake disaster anticipated through the earthquake prediction information, which is listed under Paragraph 2-(1) ~ (3) of the same Article.

Organization and Task of the Municipal Headquarters

Article 18. The municipal headquarters shall be headed by a general director, who shall be the mayor of the municipality.

2. The municipal headquarters shall perform the tasks listed below:

1) To implement short-term prevention measures against earthquake disaster, etc. in the municipal area and to promote the implementation.

2) Matters within the power of the general director of the municipal headquarters prescribed under the next Paragraph.

3) In addition to those listed above, those matters within its power as prescribed under laws and cabinet orders based on them.

3. In order to carry out short-term prevention measures against earthquake disaster, etc. in the municipal area, the general director of the municipal headquarters may, to the extent of the necessity, give necessary instructions to the board of education of the municipality concerned.

4. In addition to those prescribed above, the organization of the municipal headquarters and other necessary matters may be prescribed under the municipal by laws.

Abolition of Prefectural Headquarters or Municipal Headquarters

Article 19. The prefectural headquarters or the municipal headquarters shall be abolished with the establishment of headquarters for disaster control as prescribed under Article 23, Paragraph 1 of the Disaster Countermeasures Basic Law regarding the earthquake disaster anticipated through the earthquake prediction information. **2.** The prefectural headquarters or the municipal headquarters shall be abolished promptly with the cancellation of the earthquake warnings statement prescribed under Article 9, Paragraph 3.

Corresponding Application of the Disaster Countermeasures Basic Law regarding Transmission of Earthquake Prediction Information

Article 20. The provision under Article 51 of the Disaster Countermeasures Basic Law shall be applied correspondingly regarding the transmission of earthquake prediction information, the provision under Article 52 regarding disaster prevention signals when an earthquake warnings statement is issued and the provisions under Article 55-57 when the prefectural governors or mayors of municipalities are informed of the earthquake warnings statement. In this case, "and public bodies and administrators of facilities important in disaster prevention (hereafter referred to as 'those responsible for emergency measures against disaster' in Article 58)" shall be understood as "and those responsible for the implementation of short-term prevention measures against earthquake disaster prescribed under Article 2-(14) of the Large-Scale Earthquake Countermeasures Act".

Short-Term Prevention Measures Against Earthquake Disaster and the Responsibility for Their Implementation

Article 21. Short-term prevention measures against earthquake disaster shall be taken regarding the matters listed below:
1) Transmission of earthquake prediction information and recommendations or instructions of evacuation.
2) Emergency measures regarding fire-fighting, flood prevention and others.
3) Aid to those who are in need of immediate assistance and other matters regarding protection.
4) Maintenance and inspection of facilities and equipments.
5) Crime prevention, traffic control and other matters regarding the preservation of social order in the area which is likely to meet disaster by the large-scale earthquake.
6) To ensure emergency transportation.
7) Arrangement of a necessary system to carry out emergency measures to secure food, medical and other supplies, public health such as cleaning and prevention of infectious diseases and other matters when an earth-

quake disaster should occur.

8) In addition to those listed above, emergency measures to prevent earth-quake disaster or to reduce the damage.

2. When an earthquake warnings statement is issued, chief officers of designated administrative organs and designated local administrative organs, heads of local public bodies and other executive organs, designated public corporations, those who formulated short-term plan of earthquake disaster prevention and others responsible for the implementation of short-term prevention measures against earthquake disaster prescribed under legal provisions shall carry out short-term prevention measures against earthquake disaster as prescribed under legal provisions or earthquake disaster prevention plans.

3. Those prescribed above shall cooperate with one another to carry out short-term prevention measures against earthquake disaster appropriately and smoothly.

Responsibilities of the Citizens

Article 22. When an earthquake warnings statement is issued, residents within the intensified area are urged to take necessary measures regarding self-control of the use of fire, driving, dangerous work and others, preparation for fire-fighting and others to prevent or to reduce the disaster. At the same time, they are urged to cooperate in those short-term prevention measures against earthquake disaster carried out by the mayors of municipalities, police officers, maritime safety officials and others.

Instructions from Mayors of Municipalities

Article 23. When those who notified as prescribed under the provisions of Article 7, Paragraph 6 or Article 8, Paragraph 2 (excluding those prescribed by the cabinet order) are found to be not carrying out short-term prevention measures against earthquake disaster prescribed under the provision of Article 21, Paragraph 2, after the issuance of earthquake warnings statement, mayors of municipalities may instruct them to carry out the measures immediately.

2. When mayors of municipalities recognize that those facilities or enterprises administered or operated by those prescribed under the provisions of Article 7, Paragraph 1 or Paragraph 2 but not notified as prescribed under the provisions of Paragraph 6 of the same Article or Article 8, Paragraph 2 (excluding those prescribed by the cabinet order) are likely to cause a dangerous situation if the earthquake should occur after the issuance of earthquake warnings statement, mayors of municipalities may instruct them to carry out necessary measures immediately to prevent the occurrence of the dangerous situation.

3. When an earthquake warnings statement is issued, mayors of municipalities may instruct occupants, owners or administrators of objects which are likely to cause a dangerous situation if the earthquake should occur (excluding those prescribed under the provision of Article 6, Paragraph 1 or Article 7, Paragraph 1 or Paragraph 2) to

remove the objects concerned or take necessary safety measures immediately, to the extent of the necessity, in order to prevent or reduce the earthquake disaster.

4. In addition to those prescribed above, when mayors of municipalities recognize that it is necessary to prevent or reduce the disaster which may be caused by the earthquake after the issuance of earthquake warnings statement, they may request or recommend those prescribed under the above three Paragraphs to take necessary measures.

5. Prefectural governors, directors of police headquarters or chiefs of offices of regional maritime safety headquarters prescribed by the cabinet order may, upon request from mayors of municipalities, instruct, request or recommend as prescribed under the above Paragraphs.

Ban or Restrictions on Traffic

Article 24. When prefectural public safety commissions of the prefectures in the intensified area or those of the adjacent prefectures recognize that it is necessary to carry out evacuation of residents, sojourners and others within the intensified area smoothly, or that it is necessary to secure emergency transportation to carry out short-term prevention measures against earthquake disaster such as emergency transportation of personnel engaged in short-term prevention measures or supplies necessary for the short-term measures after the issuance of earthquake warnings statement, they may ban or restrict pedestrian or vehicle traffic to the extent of the necessity as prescribed by the cabinet order.

Warnings, Instructions and Others by Police Officers during Evacuation

Article 25. When police officers recognize that a dangerous situation is likely to occur with congestion due to evacuation after the issuance of earthquake warnings statement, they may give necessary warnings or instructions to those who may cause danger or who may be in danger in order to prevent the occurrence of the dangerous situation. In this case, police officers may, if it is found to be especially necessary, take necessary measures such as prohibiting entry into dangerous places, withdrawal of persons from such places, or removal of vehicles and other objects on the road which may cause danger.

Corresponding Application of the Disaster Countermeasures Basic Law in Respect of Short-Term Prevention Measures against Earthquake Disaster

Article 26. The provisions under Articles 58, 60, 61, 63, 67, 68, 74 and 79 of the Disaster Countermeasures Basic Law shall be applied correspondingly in the situation when earthquake warnings statement is issued. In this case, "those responsible for emergency measures against disaster" in Article 58 shall be understood as "those responsible for the implementation of short-term prevention measures against earthquake disaster prescribed under Article 2-(14) of the Large-Scale Earthquake Countermeasures Act," and "report" in Article 60, Paragraph 3 as "report and notify the chief of police station concerned."

2. The provision under Article 72 of the Disaster Countermeasures Basic Law shall be applied correspondingly to the instructions which prefectural governors give to mayors of municipalities when earthquake warnings statement is issued.

3. The provision under Article 86 of the Disaster Countermeasures Basic Law shall be applied correspondingly to lease or use of properties of State or local public bodies for the implementation of short-term prevention measures against earthquake disaster.

Special Cases of Public Use of Private Properties in Emergency

Article 27. When mayors of municipalities recognize that there is an urgent need for the implementation of short-term prevention measures against earthquake disaster, they may use any plot, buildings or any other structures belonging to any person within the municipalities concerned temporarily or use soil and stone, bamboo and lumber, and other objects as prescribed by the cabinet order.

2. The provision under Article 63, Paragraph 2 of the Disaster Countermeasures Basic Law shall be applied correspondingly to the case of the preceding Paragraph.

3. Prefectural governors may, if it is found to be especially necessary for the implementation of short-term prevention measures against earthquake disaster regarding those matters mentioned under Article 21, Paragraph 1-(4) ~ (8), invoke the provisions of Articles 25 ~ 27 of the Disaster Relief Law (Law No. 118, 1947) and issue orders for cooperation or custody, use plots, houses and materials, expropriate materials, cause their officials to enter and inspect the place where the materials are located or stored, or collect necessary information from the person who had the materials stored.

4. The power of prefectural governors prescribed under the preceding Paragraph may, as prescribed by the cabinet order, be delegated partly to the mayors of municipalities.

5. Chief officers of designated administrative organs and local administrative organs may, if it is found to be especially necessary for the implementation of short-term prevention measures against earthquake disaster regarding matters listed under Article 21, Paragraph 1-(4) ~ (8), as prescribed under the intensified plan of earthquake disaster prevention, order those who are engaged in production, collection, sales, distribution, storage or transportation of materials necessary for carrying out the measures concerned to store the materials they handle, or cause their officials to enter and inspect the place where the materials are located or stored, or collect necessary information from the person who had the materials stored.

6. The State or local public bodies shall, if the actions prescribed under Paragraphs 1, 3 and the preceding Paragraph have been taken, compensate for any normal loss that may result from the actions.

7. With regard to the actions prescribed under Paragraph 3 or 5, prefectural governors, mayors of municipalities, chief officers of designated administrative organs and designated local administrative organs shall serve official orders before taking the actions, as prescribed by the cabinet order.

8. The official order prescribed above shall have the following matters entered as prescribed by the cabinet order.

1) Name and address of the person who is served with the official order (in the case of a corporation, the name and address of its main office).

2) The provisions of laws on which the action is based.

3) In the case of an order for custody, kind, quantity, place of custody and duration in respect of the materials to be kept. For the use of plot or a house, its location and the duration of such use. For the use or appropriation of materials, kind, quantity, location, duration of such use or the date of appropriation.

9. The provision under Article 83 of the Disaster Countermeasures Basic Law is applied correspondingly to the case of entry of prefectural officials as prescribed under the provision of Paragraph 3, and also to the case of entry of officials of designated administrative organs or designated local administrative organs as prescribed under the provision of Paragraph 5.

Report on the Situation regarding Evacuation and Others

Article 28. When an earthquake warnings statement is issued, mayors of municipalities shall, as prescribed by the cabinet order, report to the prefectural headquarters on the situation regarding evacuation and others. In this case, the general director of the prefectural headquarters shall give an outline of the report to the Headquarters.

2. Mayors of municipalities shall report to the prefectural headquarters on the situation regarding the short-term prevention measures against earthquake disaster being taken, as prescribed by the cabinet order. Chief officers of designated administrative organs, representatives of designated public corporations, general directors of prefectural headquarters, and the general director of the headquarters of disaster prevention for petroleum complex and others shall report to the Headquarters on the situation regarding the measures being taken, as prescribed by the cabinet order.

Financial Assistance

Article 29. For smooth implementation of the work regarding those facilities which shall be provided promptly, based on the intensified plan of earthquake disaster prevention, the State may, within the limits of its budget, subsidize part of the expenses required for the work, or take other necessary measures.

Bearing of Expenses Required for Short-Term Prevention Measures against Earthquake Disaster

Article 30. Except for the cases that there are special legal provisions or special measures are being taken within the limits of the budget, expenses required for short-term prevention measures against earthquake disaster and other expenses required for the enforcement of this Act shall be borne by those responsible for the implementation.

Corresponding Application of the Disaster Countermeasures Basic Law in Respect of Financial Measures

Article 31. The provision under Article 92 of the Disaster Countermeasures Basic Law shall be applied correspondingly to the expenses required for support prescribed under Article 67, Paragraph 1, Article 68, Paragraph 1 or Article 74, Paragraph 1 of the same Law, which are applied correspondingly in Article 26, Paragraph 1 of this Act: the provision under Article 93 of the same Law to the expenses required for both short-term prevention measures against earthquake disaster and support carried out by mayors of municipalities as directed by prefectural governors under the provision of Article 72 of the Law which is applied correspondingly in Article 26, Paragraph 2; the provision under Article 94 of the Law to the expenses required for short-term prevention measures against earthquake disasters and the provision under Article 95 of the Law to the expenses required for short-term prevention measures against earthquake disaster carried out by heads of local public bodies as directed by the General Director of the National Headquarters for Earthquake Disaster Prevention under the provision of Article 13, Paragraph 1.

Earthquake Disaster Prevention Exercises in Intensified Areas

Article 32. When an intensified area is designated under the provision of Article 3, Paragraph 1, chief officers of designated administrative organs and designated local administrative organs, heads of local public bodies and other executive organs, designated public corporations, those who formulated short-term plan of earthquake disaster prevention and those responsible for the implementation of the short-term prevention measures against earthquake disaster prescribed under legal provisions shall, either individually or in cooperation, carry out earthquake disaster prevention exercises as prescribed by legal provisions or by earthquake disaster prevention plans.
2. For effective implementation of the earthquake disaster prevention exercises prescribed above, the prefectural public safety commissions may, if it is found to be especially necessary, designate section of the road and ban or restrict pedestrian or vehicle traffic to the extent necessary for the exercises as prescribed by the cabinet order.
3. Those prescribed under Paragraph 1 may, for the purpose of carrying out disaster

prevention exercises prescribed under the Paragraph, request cooperation from the residents, public and private bodies concerned.

Promotion of Science and Technology, etc.

Article 33.   In order to predict the occurrence of earthquakes, the State shall make efforts to improve the facilities and equipments for seismological observations and surveys. At the same time, the State shall endeavor to improve the system of research, promote research and diffuse its results for the advancement of science and technology which will contribute to earthquake prediction.

Application of This Act to Special Wards

Article 34.   In the application of this Act, special wards are regarded as cities.

Delegation to the Cabinet Order

Article 35.   Except those prescribed under this Act, the procedure for the implementation of this Act and other necessary matters regarding the enforcement of this Act shall be prescribed by the cabinet order.

Penalties

Article 36.   Those who come under one of the categories below shall be sentenced to six months' imprisonment with labor or a fine of not exceeding ￥200,000.
   1)   Those who did not follow orders for cooperation or custody from prefectural governors (including mayors of municipalities who have been delegated the power as prescribed under Article 27 Paragraph 4) as prescribed under Article 27, Paragraph 3.
   2)   Those who did not follow orders for custody from chief officers or designated administrative organs or designated local administrative organs (including those officials who have been delegated the power under the provision of Article 27, Paragraph 1 of the Disaster Countermeasures Basic Law, applied correspondingly in Article 15) as prescribed under the provision of Article 27, Paragraph 5.

Article 37.   Those drivers of vehicles, who did not observe traffic ban of restrictions by the prefectural public safety commissions as prescribed under the provision of Article 24, shall be sentenced to three months' imprisonment with labor or a fine of not exceeding ￥100,000.

Article 38.   Those who come under one of the categories below shall be sentenced to a fine of not exceeding ￥100,000.
   1)   Those who refused, obstructed or avoided spot inspection prescribed under the provision of Article 27, Paragraph 3 (including those cases of

delegation of power prescribed under the provision of Article 27, Paragraph 4; applicable hereunder) or Paragraph 5 (including those cases of delegation of power prescribed under the provision of Article 27, Paragraph 1 of the Disaster Countermeasures Basic Law applied correspondingly in Article 15; applicable hereunder).

2) Those who did not report as prescribed under the provision of Article 27, Paragraph 3 or Paragraph 5, or who submitted false reports.

Article 39. Those who come under one of the categories below shall be sentenced to a fine of not exceeding ¥50,000 or penal detention.

1) Those who used disaster prevention signals unlawfully, which are prescribed by an order of the Prime Minister's Office based on the provision under Article 52, Paragraph 1 of the Disaster Countermeasures Basic Law applied correspondingly in Article 20, or those who used similar signals.

2) Those who did not follow prohibition, restriction or orders for withdrawal by mayors of municipalities under the provision of Article 63, Paragraph 1 of the Disaster Countermeasures Basic Law applied correspondingly in Article 26, Paragraph 1, or by police officers or maritime safety officials under the provision of Article 63, Paragraph 2 of the same Law.

Article 40. If representatives of corporations or authorized agents, employees and other workers of corporations or individuals violate Article 36 or Article 38 in respect of the affairs of the corporations or the individuals, they shall be punished. In addition, the corporations or the individuals shall also be fined as prescribed under the provisions of Articles concerned.

APPENDIX II:

EARTHQUAKE HAZARDS REDUCTION ACT OF 1977

91 STAT. 1098   PUBLIC LAW 95–124—OCT. 7, 1977

Public Law 95–124
95th Congress

An Act

Oct. 7, 1977
[S. 126]
To reduce the hazards of earthquakes, and for other purposes.

Be it enacted by the Senate and House of Representatives of the United States of America in Congress assembled.

Earthquake
Hazards
Reduction Act of
1977.
42 USC 7701
note
42 USC 7701

Sec. 1.   Short Title
That this Act may be cited as the "Earthquake Hazards Reduction Act of 1977."

Sec. 2.   Findings
The Congress finds and declares the following:
(1) All 50 States are vulnerable to the hazards of earthquakes, and at least 39 of them are subject to major or moderate seismic risk, including Alaska, California, Hawaii, Illinois, Massachusetts, Missouri, Montana, Nevada, New Jersey, New York, South Carolina, Utah, and Washington. A large portion of the population of the United States lives in areas vulnerable to earthquake hazards.
(2) Earthquakes have caused, and can cause in the

future, enormous loss of life, injury, destruction of property, and economic and social disruption. With respect to future earthquakes, such loss, destruction, and disruption can be substantially reduced through the development and implementation of earthquake hazards reduction measures, including (A) improved design and construction methods and practices, (B) land-use controls and redevelopment, (C) prediction techniques and early-warning systems, (D) coordinated emergency preparedness plans, and (E) public education and involvement programs.

(3) An expertly staffed and adequately financed earthquake hazards reduction program, based on Federal, State, local, and private research, planning, decisionmaking, and contributions would reduce the risk of such loss, destruction, and disruption in seismic areas by an amount far greater than the cost of such program.

(4) A well-funded seismological research program in earthquake prediction could provide data adequate for the design, of an operational system that could predict accurately the time, place, magnitude, and physical effects of earthquakes in selected areas of the United States.

(5) An operational earthquake prediction system can produce significant social, economic, legal, and political consequences.

(6) There is a scientific basis for hypothesizing that major earthquakes may be moderated, in at least some seismic areas, by application of the findings of earthquake control and seismological research.

(7) The implementation of earthquake hazards reduction measures would, as an added benefit, also reduce the risk of loss, destruction, and disruption from other natural hazards and manmade hazards, including hurricanes, tornadoes, accidents, explosions, landslides, building and structural cave-ins, and fires.

(8) Reduction of loss, destruction, and disruption from earthquakes will depend on the actions of individuals, and organizations in the private sector and governmental units at Federal, State, and local levels. The current capability to transfer knowledge and information to these sectors is insufficient. Improved mechanisms are needed to translate existing information and research findings into reasonable and usable specifications, criteria, and practices so that individuals, organizations, and governmental units may

make informed decisions and take appropriate actions.

(9)   Severe earthquakes are a worldwide problem. Since damaging earthquakes occur infrequently in any one nation, international cooperation is desirable for mutual learning from limited experiences.

(10)   An effective Federal program in earthquake hazards reduction will require input from and review by persons outside the Federal Government expert in the sciences of earthquake hazards reduction and in the practical application of earthquake hazards reduction measures.

42 USC 7702.

Sec. 3.   Purpose

It is the purpose of the Congress in this Act to reduce the risks of life and property from future earthquakes in the United States through the establishment and maintenance of an effective earthquake hazards reduction program.

42 USC 7703.

Sec. 4.   Definitions

As used in this Act, unless the context otherwise requires:

(1)   The term "includes" and variants thereof should be read as if the phrase "but is not limited to" were also set forth.

(2)   The term "program" means the earthquake hazards reduction program established under section 5.

(3)   The term "seismic" and variants thereof mean having to do with, or caused by earthquakes.

(4)   The term "State" means each of the States of the United States, the District of Columbia, the Commonwealth of Puerto Rico, the Virgin Islands, Guam, American Samoa, the Commonwealth of the Mariana Islands, and any other territory or possession of the United States.

(5)   The term "United States" means, when used in a geographical sense, all of the States as difined in section 4(4).

42 USC 7704.

Sec. 5.   National Earthquake Hazards Reduction Program

(a)   ESTABLISHMENT.—The President shall establish and maintain, in accordance with the provisions and policy of this Act, a coordinated earthquake hazards reduction program, which shall—

(1)   be designed and administered to achieve the objectives set forth in subsection (c);

(2) involve, where appropriate, each of the agencies listed in subsection (d); and

(3) include each of the elements described in subsection (e), the implementation plan described in subsection (f), and the assistance to the States specified in subsection (g).

(b) DUTIES.—The President shall—

(1) within 30 days after the date of enactment of this Act, designate the Federal department, agency, or entity responsible for the development of the implementation plan described in subsection (f);

(2) within 210 days after such date of enactment, submit to the appropriate authorizing committees of the Congress the implementation plan described in subsection (f); and

(3) by rule, within 300 days after such date of enactment—

(A) designate the Federal department, agency, or interagency group which shall have primary responsibility for the development and implementation of the earthquake hazards reduction program;

(B) assign and specify the role and responsibility of each appropriate Federal department, agency, and entity with respect to each object and element of the program;

(C) establish goals, priorities, and target dates for implementation of the program;

(D) provide a method for cooperation and coordination with, and assistance (to the extent of available resources) to, interested governmental entities in all States, particularly those containing areas of high or moderate seismic risk; and

(E) provide for qualified staffing for the program and its components.

(c) OBJECTIVES.—The objectives of the earthquake hazards reduction program shall include—

(1) the development of technologically and economically feasible design and construction methods and procedures to make new and existing structures, in areas of seismic risk, earthquake resistant, giving priority to the development of such methods and procedures for nuclear power generating plants, dams, hospitals, schools, public utilities, public safety structures, high occupancy buildings, and other structures which are especially needed in time of disaster;

Earthquake
prediction.

(2)   the implementation in all areas of high or moderate seismic risk, of a system (including personnel, technology, and procedures) for predicting damaging earthquakes and for identifying, evaluating, and accurately characterizing seismic hazards;

Model codes.

(3)   the development, publication, and promotion, in conjunction with State and local officials and professional organizations, of model codes and other means to coordinate information about seismic risk with land-use policy decisions and building activity;

Earthquake-
related issues,
understanding.

(4)   the development, in areas of seismic risk, of improved understanding of, and capability with respect to, earthquake-related issues, including methods of controlling the risks from earthquakes, planning to prevent such risks, disseminating warnings of earthquakes, organizing emergency services, and planning for reconstruction and redevelopment after an earthquake;

(5)   the education of the public, including State and local officials, as to earthquake phenomena, the identification of locations and structures which are especially susceptible to earthquake damage, ways to reduce the adverse consequences of an earthquake, and related matters;

Research.

(6)   the development of research on—

(A)   ways to increase the use of existing scientific and engineering knowledge to mitigate earthquake hazards;

(B)   the social, economic, legal, and political consequences of earthquake prediction; and

(C)   ways to assure the availability of earthquake insurance or some functional substitute; and

(7)   the development of basic and applied research leading to a better understanding of the control or alteration of seismic phenomena.

(d)   PARTICIPATION.—In assigning the role and responsibility of Federal departments, agencies, and entities under subsection (b) (3) (B), the President shall, where appropriate, include the United States Geological Survey, the National Science Foundation, the Department of Defense, the Department of Housing and Urban Development, the National Aeronautics and Space Administration, the National Oceanic and Atmospheric Administration, the National Bureau of Standards, the Energy Research and Development Administration, the Nuclear Regulatory Commission, and the National Fire

Prevention and Control Administration.

(e) RESEARCH ELEMENTS.—The research elements of the program shall include—

(1) research into the basic causes and mechanisms of earthquakes;

(2) development of methods to predict the time, place, and magnitude of future earthquakes;

(3) development of an understanding of the circumstances in which earthquakes might be artificially induced by the injection of fluids in deep wells, by the impoundment of reservoirs, or by other means;

(4) evaluation of methods that may lead to the development of a capability to modify control earthquakes in certain regions;

(5) development of information and guidelines for zoning land in light of seismic risk in all parts of the United States and preparation of seismic risk analyses useful for emergency planning and community preparedness;

(6) development of techniques for the delineation and evaluation of the political effects of earthquakes, and their application on a regional basis;

(7) development of methods for planning, design, construction, rehabilitation, and utilization of manmade works so as to effectively resist the hazards imposed by earthquakes;

(8) exploration of possible social and economic adjustments that could be made to reduce earthquake vulnerability and to exploit effectively existing and developing earthquake mitigation techniques; and

(9) studies of foreign experience with all aspects of earthquakes.

(f) IMPLEMENTATION PLAN.—The President shall develop, through the Federal agency, department, or entity designated under subsection (b) (1), an implementation plan which shall set year-by-year targets through at least 1980, and shall specify the roles for Federal agencies, and recommended appropriate roles for State and local units of government, individuals, and private organizations, in carrying out the implementation plan. The plan shall provide for—

(1) the development of measures to be taken with respect to preparing for earthquakes, evaluation of prediction techniques and actual predictions of earthquakes, warning the residents of an area that an earthquake may

occur, and ensuring that a comprehensive response is made
to the occurrence of an earthquake;

(2)   the development of ways for State, county, local,
and regional governmental units to use existing and developing knowledge about the regional and local variations
of seismic risk in making their land use decisions;

(3)   the development and promulgation of specifications,
building standards, design criteria, and construction practices to achieve appropriate earthquake resistance for new
and existing structures;

(4)   an examination of alternative provisions and re-
quirements for reducing earthquake hazards through
Federal and federally financed construction, loan guarantees, and licenses;

(5)   the determination of the appropriate role for insurance, loan programs, and public and private relief efforts in
moderating the impact of earthquakes; and

(6)   dissemination, on a timely basis, of—

(A)   instrument-derived data of interest to other
researchers;

(B)   design and analysis data and procedures of
interest to the design professions and to the construction industry; and

(C)   other information and knowledge of interest to
the public to reduce vulnerability to earthquake
hazards.

Report, filing
with
congressional
committees.

When the implementation plan developed by the President under
this section contemplates or proposes specific action to be taken
by any Federal agency, department, or entity, and, at the end of
the 30-day period beginning on the date the President submits
such plan to the appropriate authorizing committees of the Congress any such action has not been initiated, the President shall
file with such committees a report explaining, in detail, the
reasons why such action has not been initiated.

(g)   STATE ASSISTANCE.—In making assistance available to
the States under the Disaster Relief Act of 1974 (42 U.S.C. 5121
et seq.), the President may make such assistance available to
further the purposes of this Act, including making available to
the States the results of research and other activities conducted
under this Act.

(h)   PARTICIPATION.—In carrying out the provisions of this
section, the President shall provide an opportunity for participation by the appropriate representatives of State and local

governments, and by the public, including representatives of business and industry, the design professions, and the research community, in the formulation and implementation of the program.

Program plan review.
Report to Congress.

Such non-Federal participation shall include periodic review of the program plan, considered in its entirety, by an assembled and adequately staffed group of such representatives. Any comments on the program upon which such group agrees shall be reported to the Congress.

Measures developed pursuant to paragraph 5 (f) (1) for the evaluation of prediction techniques and actual predictions of earthquakes shall provide for adequate non-Federal participation. To the extent that such measures include evaluation by Federal employees of non-Federal prediction activities, such measures shall also include evaluation by persons not in full-time Federal employment of Federal prediction activities.

42 USC 7705.
Submittal to congressional committees.

Sec. 6. Annual Report

The President shall, within ninety days after the end of each fiscal year, submit an annual report to the appropriate authorizing committees in the Congress describing the status of the program, and describing and evaluating progress achieved during the preceding fiscal year in reducing the risks of earthquake hazards. Each such report shall include any recommendations for legislative and other action the President deems necessary and appropriate.

42 USC 7706.

Sec. 7. Authorization of Appropriations

(a) GENERAL.—There are authorized to be appropriated to the President to carry out the provisions of sections 5 and 6 of this Act (in addition to any authorizations for similar purposes included in other Acts and the authorizations set forth in subsections (b) and (c) of this section), not to exceed $1,000,000 for the fiscal year ending September 30, 1978, not to exceed $2,000,000 for the fiscal year ending September 30, 1979, and not to exceed $2,000,000 for the fiscal year ending September 30, 1980.

(b) GEOLOGICAL SURVEY.—There are authorized to be appropriated to the Secretary of the Interior for purposes for carrying out, through the Director of the United States Geological Survey, the responsibilities that may be assigned to the Director under this Act not to exceed $27,500,000 for the fiscal year ending September 30, 1978; not to exceed $35,000,000 for the fiscal year ending September 30, 1979; and not to exceed $40,000,000 for the fiscal year ending September 30, 1980.

(c) NATIONAL SCIENCE FOUNDATION.—To enable the Foundation to carry out responsibilities that may be assigned to it under this Act, there are authorized to be appropriated to the Foundation not to exceed $27,500,000 for the fiscal year ending September 30, 1978; not to exceed $35,000,000 for the fiscal year ending September 30, 1979; and not to exceed $40,000,000 for the fiscal year ending September 30, 1980.

Approved October 7, 1977.

REFERENCES

ACADEMIA SINICA, SEISMOLOGICAL COMMITTEE, *Chronological Tables of Earthquake Data of China*, 2 Vol., 1653 pp., Science Press, Peking, 1956 (in Chinese).

ACADEMIA SINICA, GEOPHYSICAL INSTITUTE, *Catalogue of Chinese Earthquakes*, 2 Vol., 361 and 302 p., Science Press, Peking, 1971 (in Chinese).

ACADEMIA SINICA, GEOPHYSICAL INSTITUTE, *Summary of Large Earthquakes in China from 780 B.C. to 1976 with Magnitude* ≥ 6, 29 pp., Map Press, Peking, 1976a (in Chinese).

ACADEMIA SINICA, GEOPHYSICAL INSTITUTE, *Distribution Map of Strong Earthquakes in China*, Map Press, Peking, 1976b.

ADAMS, R., The Haicheng, China, earthquake of 4 February, 1975, The first successful predicted major earthquake, *Earthq. Eng. Struct. Dyn.*, **4**, 423–437, 1975.

AGGARWAL, Y. P., L. R. SYKES, D. W. SIMPSON, and P. G. RICHARDS, Spatial and temporal variations in t_s/t_p and in P-wave residuals at Blue Mountain Lake, New York; application to earthquake prediction, *J. Geophys. Res.*, **80**, 718–732, 1975.

ALLEN, C., M. G. BONILLA, W. F. BRACE, M. BULLOCK, R. W. CLOUGH, R. M. HAMILTON, R. HOFHEINZ, Jr., C. KISSLINGER, L. KNOPOFF, M. PARK, F. PRESS, C. B. RALEIGH, and L. R. SYKES, Earthquake research in China, *EOS (Trans. Am. Geophys. Union)*, **55**, 838–882, 1975.

AMBRASEYS, N. N., Some characteristic features of the Anatolian fault zone, *Tectonophysics*, **9**, 143–165, 1970.

ANDO, M., A fault-origin model of the great Kanto earthquake of 1923 as deduced from geodetic data, *Bull. Earthq. Res. Inst., Univ. Tokyo*, **49**, 19–32, 1971.

ANDO, M., Seismotectonics of the 1923 Kanto earthquake, *J. Phys. Earth*, **22**, 263–277, 1974.

ASIMOV, M. S., J. S. ERJANOV, K. E. KALMURSAIEV, M. K. KURGANOV, F. A. MAVIJANOV, S. H. NEGMATULAEV, I. L. NERSESOV, and V. I. OULOMOV, On the state of the research concerning earthquake prediction in the Soviet Republics of Central Asia, Presented at the 1979 UNESCO Symposium on Earthquake Prediction, 1979 (in Russian).

BARSUKOV, O. M., Variations of electric resistivity of mountain rocks connected with tectonic causes, *Tectonophysics*, **14**, 273–277, 1972.

BOLT, B. A., Earthquake studies in the People's Republic of China, *EOS (Trans. Am. Geophys. Union)*, **55**, 108–117, 1974.

BRADY, B. T., Theory of earthquakes, 1. A scale independent theory of rock failure, *Pure Appl. Geophys.*, **112**, 701–725, 1974.

BUFE, C. G., W. H. BAKUN, and D. TOCHER, Geophysical studies in the San Andreas fault zone at the Stone Canyon Observatory, California, in *Proceedings of the Conference on Tectonic Problems on the San Andreas Fault System, Geol.*, Vol. 13, edited by R. L. Kovach and A. Nur, pp. 36–93, Stanford Univ. Publ., 1973.

BUFE, C. G., J. H. PFLUKE, and R. L. WESSON, Premonitory vertical migration of microearthquakes in Central California—Evidence of dilatancy biasing? *Geophys. Res. Lett.*, **1**, 221–224, 1974.

BURFORD, R. O., R. D. NASON, and P. W. HARSH, Studies of fault creep in Central California, *Earthq. Inf. Bull.*, **10**, 174–181, 1978.

CASTLE, R. O., Leveling surveys and the southern California uplift, *Earthq. Inf. Bull.*, **10**, 88–92, 1978.

CASTLE, R. O., J. P. CHURCH, and M. R. ELLIOT, Aseismic uplift in southern California, *Science*, **192**, 251–253, 1976.

CHOU, C. W. and R. S. CROSSON, Search for time-dependent seismic travel times from mining explosions near Centralia, Washington, *Geophys. Res. Lett.*, **5**, 97–100, 1978.

COE, R. S., Earthquake prediction program in the People's Republic of China, *EOS (Trans. Am. Geophys. Union)*, **52**, 940–943, 1971.

CORWIN, P. F. and H. F. MORRISON, Self-potential variations preceding earthquakes in central California, *Geophys. Res. Lett.*, **4**, 171–174, 1977.

CRAIG, H., J. E. LUPTON, Y. CHUNG, and R. M. HOROWITZ, Investigation of radon and helium as possible fluid-phase precursors to earthquakes, Technical report, No. 1 (SIO Ref. No. 75–15), 1975.

DAMBARA, T., Vertical movements of the earth's crust in relation to the Matsushiro earthquake, *J. Geod. Soc. Japan*, **12**, 18–45, 1966 (in Japanese).

DAMBARA, T., Crustal movements before, at and after the Niigata earthquake, *Rep. Coord. Comm. Earthq. Predict.*, **9**, 93–96, 1973 (in Japanese).

DAMBARA, T., A revised relation between the area of the crustal deformation associated with an earthquake and its magnitude, *Rep. Coord. Comm. Earthq. Predict.*, **21**, 167–169, 1979 (in Japanese).

DAMBARA, T., Geodesy and earthquake prediction, in *Current Research in Earthquake Prediction I.*, edited by T. Rikitake, Center for Academic Publications Japan/D. Reidel Publishing Company, Tokyo, 167–220, 1981.

DICKSON, D., Budget cuts cause upheavals in earthquake research, *Nature*, **277**, 421, 1979.

DUAN, Xing-bei, Jian-zhong ZHENG, Zhi-gun ZHOU, Shou-min YAN, and Ci-chang SUN, Variations in teleseismic P wave residuals before the Haicheng earthquake, *Acta Geophys. Sinica*, **19**, 286–294, 1976 (in Chinese).

EARTHQUAKE DISASTER PREVENTION COMMITTEE, *Zotei Dai Nihon Jishin Shiryo (Supplement to Dai Nihon Jishin Shiryo)*, Vol. 1, 945 pp., Vol. 2, 754 pp., Vol. 3, 945 pp., Tokyo, 1941–1943, Reprinted by Meiho Sha, Tokyo, 1975 and 1976 (in Japanese).

ENGDAHL, E. R. and C. KISSLINGER, Seismological precursors to a magnitude 5 earthquake in the central Aleutian Islands, *J. Phys. Earth*, **25**, Suppl., S243–S250, 1977.

EVERNDEN, J. F. (Convener), *Proceedings of Conference I Abnormal Animal Behavior Prior to Earthquakes*, I., 429 pp., U.S. Geological Survey, Menlo Park, California, 1976.

EVERNDEN, J. F. (Convener), B. L. ISACKS, and G. PLAFKER (Organizers), *Proceedings of Conference VI Methodology for Identifying Seismic Gaps*, 924 pp., U.S. Geological Survey, Openfile Rep. 78–943, Menlo Park, California, 1978.

EVISON, F. F., Fluctuations of seismicity before major earthquakes, *Nature*, **266**, 710–712, 1977a.

EVISON, F. F., The precursory earthquake swarm, *Phys. Earth Planet. Inter.*, **15**, 19–23, 1977b.

FEDOTOV, S. A., Regularities of the distribution of strong earthquakes in Kamchatka, the Kurile Islands, and northeastern Japan, *Trudy Inst. Phys. Earth, Acad. Sci. U.S.S.R.*, No. 36 (203), 66–93, 1965.

FEDOTOV, S. A., N. A. DOLBILKINA, V. N. MOROZOV, V. I. MYACHKIN, V. B. PREOBRAZENSKY, and G. A. SOBOLEV, Investigation on earthquake prediction in Kamchatka, *Tectonophysics*, **9**, 249–258, 1970.

FEDOTOV, S. A., A. A. GUSEV, and S. A. BOLDYREV, Progress of earthquake prediction in Kamchatka, *Tectonophysics*, **14**, 279–286, 1972.

FENG, De-yi, Anomalous variations of seismic velocity ratio before the Yongshan-Daguan earthquake ($M = 7.1$) on May 11, 1974, *Acta Geophys. Sinica*, **18**, 235–239, 1975 (in Chinese).

FENG, De-yi, Ai-na TAN, and Ke-fean WANG, Velocity anomalies of seismic waves from near earthquakes and earthquake prediction, *Acta Geophys. Sinica*, **17**, 84–98, 1974 (in Chinese).

FENG, De-yi, Si-hua ZHENG, Guo-ying SHENG, Zheng-xiang FU, Shi-lei GAO, Rui-ming LUO, and Bing-can LI, Preliminary study of the velocity anomalies of seismic waves before and after some strong and moderate earthquakes in Western China, (1) The velocity ratio anomalies, *Acta Geophys. Sinica*, **19**, 196–205, 1976a (in Chinese).

FENG, Rui, Qing-yan PANG, Zheng-xiang FU, Jian-zhong ZHENG, Ci-chang SUN, and Bao-xiang LI, Variations of V_P/V_S before and after the Haicheng earthquake of 1975, *Acta Geophys. Sinica*, **19**, 295–305, 1976b (in Chinese).

FITTERMAN, D. V. and T. R. MADDEN, Resistivity observations during creep events at Melendy Ranch, California, *J. Geophys. Res.*, **82**, 5401–5408, 1977.

GARZA, T. and C. LOMNITZ, The Oaxaca gap: A case history, in *Proceedings of Conference VI Methodology for Identifying Seismic Gaps*, J. F. Evernden (Convener), B. L. Isacks and G. Plafker (Organizers), pp. 173–188, U.S. Geol. Surv., Open-file Rep. 78–943, 1978.

GELFAND, I. M., Sh. A. GUBERMAN, M. L. IZVEKOVA, V. I. KEILIS-BOROK, and E. La. RANZMAN, Criteria of high seismicity, determined by pattern recognition, *Tectonophysics*, **13**, 415–422, 1972.

GELFAND, I. M., Sh. A. GUBERMAN, V. I. KEILIS-BOROK, L. KNOPOFF, F. PRESS, E. Ya. RANZMAN, I. M. ROTWAIN, and A. M. SADOVSKY, Pattern recognition applied to earthquake epicenters in California, *Phys. Earth Planet. Inter.*, **11**, 227–283, 1976.

GEODETIC SURVEY PARTY, Coseismic gravity changes during Izu-Oshima Kinkai earthquake, *Rep. Coord. Comm. Earthq. Predict.*, **20**, 56–59, 1978 (in Japanese).

374

GEOGRAPHICAL SURVEY INSTITUTE, Recent crustal movement in South Kanto district, 4, *Rep. Coord. Comm. Earthq. Predict.*, **8**, 23–26, 1972 (in Japanese).

GEOGRAPHICAL SURVEY INSTITUTE, Precise strain measurements in Yamakita and Tanna regions, *Rep. Coord. Comm. Earthq. Predict.*, **11**,94–95, 1974a (in Japanese).

GEOGRAPHICAL SURVEY INSTITUTE, Primary geodetic survey in Northeast Izu Peninsula, *Rep. Coord. Comm. Earthq. Predict.*, **12**, 49–50, 1974b (in Japanese).

GEOGRAPHICAL SURVEY INSTITUTE, G.D.P. traverse survey of high precision in Chubu and Tokai districts, *Rep. Coord. Comm. Earthq. Predict.*, **11**, 107–108, 1974c (in Japanese).

GEOGRAPHICAL SURVEY INSTITUTE, Crustal upheaval near the mouth of Tamagawa River, *Rep. Coord. Comm. Earthq. Predict.*, **13**, 34–35, 1975 (in Japanese).

GEOGRAPHICAL SURVEY INSTITUTE, Horizontal earth's strain in Suruga Bay, *Rep. Coord. Comm. Earthq. Predict.*, **17**, 146–148, 1977 (in Japanese).

GEOGRAPHICAL SURVEY INSTITUTE, Crustal deformation in Izu Peninsula, *Rep. Coord. Comm. Earthq. Predict.*, **20**, 92–99, 1978 (in Japanese).

GEOGRAPHICAL SURVEY INSTITUTE, Crustal movements in the Tokai district, *Rep. Coord. Comm. Earthq. Predict.*, **21**, 122–129, 1979a (in Japanese).

GEOGRAPHICAL SURVEY INSTITUTE, Crustal deformation in the eastern Izu district, *Rep. Coord. Comm. Earthq. Predict.*, **22**, 68–71, 1979b (in Japanese).

GEOGRAPHICAL SURVEY INSTITUTE, Horizontal crustal movement in the Tokai district, *Rep. Coord. Comm. Earthq. Predict.*, **22**, 159–162, 1979c (in Japanese).

GEOGRAPHICAL SURVEY INSTITUTE, Crustal movement in the Tokai district, *Rep. Coord. Comm. Earthq. Predict.*, **23**, 86–92, 1980a; **24**, 152–158, 1980b; **25**, 213–222, 1981 (in Japanese).

GEOLOGICAL SURVEY OF JAPAN, *In situ* stress measurements by a stress relief method in the Tanzawa Mountains, *Rep. Coord. Comm. Earthq. Predict.*, **24**, 99–103, 1980 (in Japanese).

GEOLOGICAL SURVEY OF JAPAN, *In situ* stress measurements by a stress relief method in the Tanzawa Mountains, *Rep. Coord. Comm. Earthq. Predict.*, **25**, 72–76, 1981 (in Japanese).

GEOMAGNETIC SURVEY PARTY, Repeated magnetic survey and observation of the total force intensity in the eastern part of the Izu Peninsula (2), *Rep. Coord. Comm. Earthq. Predict.*, **18**, 47–51, 1977 (in Japanese).

GEOMAGNETIC SURVEY PARTY, Repeated magnetic survey and observation of total force intensity in the eastern part of the Izu Peninsula (4), *Rep. Coord. Comm. Earthq. Predict.*, **22**, 75–78, 1979 (in Japanese).

GORDON, F. R., Water level changes preceding the Meckering, Western Australia, earthquake of October 14, 1968, *Bull. Seismol. Soc. Am.*, **60**, 1739–1740, 1970.

GREGSON, P. J., R. S. SMITH, and F. K. McCUE, An explanation of water level changes preceding the Meckering earthquake of 14 October 1968, *Bull. Seismol. Soc. Am.*, **66**, 631–632, 1976.

GUPTA, I. N., Premonitory seismic-wave phenomena before earthquakes near Fairview Peak, Nevada, *Bull. Seismol. Soc. Am.*, **65**, 425–437, 1975.

GUTENBERG, B. and C. F. RICHTER, Frequency of earthquakes in California, *Bull. Seismol. Soc. Am.*, **34**, 185–188, 1944.

HAAS, J. E. and D. S. MILETI, Consequences of earthquake prediction on other adjustments to earthquakes, Paper presented at Australian Academy of Science Symposium of National Hazards in Australia, Section 3: Community and Individual Responses to Natural Hazards, May 26–29, 1976, Canberra, Australia, 1976.

HAAS, J. E. and D. S. MILETI, Socioeconomic Impact of Earthquake Prediction on Government, Business, and Community, Institute of Behavioral Science, University of Colorado, 1977a.

HAAS, J. E. and D. S. MILETI, Socioeconomic and political consequences of earthquake prediction, *J. Phys. Earth*, **25**, Suppl., S283–S293, 1977b.

HAAS, J. E., D. S. MILETI, and J. R. HUTTON, Uses of earthquake prediction information, Paper presented at XVI General Assembly of the International Union of Geodesy and Geophysics, International Association of Seismology and Physics of the Earth's Interior, Symposium No. 4: Geophysical Phenomena Preceding, Accompanying and Following Earthquakes, Grenoble, France, September 2, 1975.

HAGIWARA, Y., Probability of earthquake occurrence as obtained from a Weibull distribution analysis of crustal strain, *Tectonophysics*, **23**, 313–318, 1974.

HAGIWARA, Y., Gravity changes associated with seismic activities, *J. Phys. Earth*, **25**, Suppl., S137–S146, 1977.

HAGIWARA, Y., H. TAJIMA, S. IZUTUYA, and H. HANADA, Gravity changes in the eastern part of Izu Peninsula during the period of 1975–1976, *J. Geod. Soc. Japan*, **22**, 201–209, 1976 (in Japanese).

HAMADA, K., Earthquake prediction research in the U.S.A. Rev. Res. Disaster Prevention, *Natl. Res. Cent. Disaster Prevent.*, **20**, 1–19, 1975 (in Japanese).

HASEGAWA, A., T. HASEGAWA, and T. HORI, Premonitory variation in seismic velocity related to the Southeastern Akita earthquake of 1970, *J. Phys. Earth*, **23**, 189–203, 1975.

HONKURA, Y., On a relation between anomalies in the geomagnetic and telluric fields observed at Nakaizu and the Izu-Oshima Kinkai earthquake of 1978, *Bull. Earthq. Res. Inst., Univ. Tokyo*, **53**, 931–937, 1978a (in Japanese).

HONKURA, Y., Electrical conductivity anomalies in the earth, *Geophys. Surv.*, **3**, 225–253, 1978b.

HONKURA, Y. and S. KOYAMA, Observations of short-period geomagnetic variations at Nakaizu (1), *Bull. Earthq. Res. Inst., Univ. Tokyo*, **53**, 925–930, 1978 (in Japanese).

HUTTON, J. R., J. H. SORENSEN and D. S. MILETI, Earthquake prediction and public reaction, in *Current Research in Earthquake Prediction I*, edited by T. Rikitake, Center for Academic Publications Japan/D. Reidel Publishing Company, Tokyo, 129–166, 1981.

IIZUKA, S., Temporal variations in V_P/V_S and some related phenomena before the 1968 Tokachi-oki earthquake, off Northeast Japan, *Zisin (J. Seismol. Soc. Japan)*, Ser. 2, **29**, 247–263, 1976a (in Japanese).

IIZUKA, S., Temporal changes in V_P/V_S ratio associated with great submarine earthquakes, off eastern Hokkaido, Japan, *Zisin (J. Seismol. Soc. Japan)*, Ser. 2, **29**, 265–275, 1976b (in Japanese).

IIZUKA, S., Temporal changes in seismic wave velocities related with the Matsushiro earthquake swarm, *Zisin (J. Seismol. Soc. Japan)*, Ser. 2, **29**, 365–374, 1976c (in Japanese).

IMPERIAL EARTHQUAKE INVESTIGATION COMMITTEE, *Dai Nihon Jishin Shiryo (Japanese Historical Records Relevant to Earthquakes)*, Part 1, 606 pp., Part 2, 595 pp., 1904, Reprinted by Shibunkan, Kyoto, 1973 (in Japanese).

INSTITUTE OF JOURNALISM, *Communication of Earthquake Information and Reaction of Local Inhabitants—Case History of the So-Called Panic Related to Aftershock Information—*, 129 pp., Unofficial Publication of the Inst. of Journalism, Univ. Tokyo, Tokyo, 1978 (in Japanese).

INSTITUTE OF JOURNALISM, *Earthquake Prediction and Social Response*, 365 pp., Inst. Journalism, Univ. Tokyo, Tokyo, 1979a (in Japanese).

INSTITUTE OF JOURNALISM, *Response to Earthquake Prediction Information*, 197 pp., Inst. Journalism, Univ. Tokyo, Tokyo, 1979b (in Japanese).

ISHIBASHI, K., Practical strategy of earthquake prediction, in *Method of Earthquake Prediction*, edited by T. Asada, pp. 193–209, Univ. Tokyo Press, Tokyo, 1978 (in Japanese).

ISHIDA, M. and H. KANAMORI, The spatio-temporal variation of seismicity before the 1971 San Fernando earthquake, California, *Geophys. Res. Lett.*, **4**, 345–346, 1977.

ISHIDA, M. and H. KANAMORI, The foreshock activities of the 1971 San Fernando earthquake, California, *Bull. Seismol. Soc. Am.*, **68**, 1265–1279, 1978.

ISHIMOTO, M. and K. IIDA, Observations sur les séismes enregistrés par le microsismographe construit dernièrement, 1, *Bull. Earthq. Res. Inst., Univ. Tokyo*, **17**, 443–478, 1939.

IZUTUYA, S., Revised results of levelling surveys during the Matsushiro earthquake swarm, *Bull. Earthq. Res. Inst., Univ. Tokyo*, **50**, 273–280, 1975 (in Japanese).

JAMISON, D. B. and N. G. W. COOK, Note on measured values for the state of stress in the earth's crust, *J. Geophys. Res.*, **85**, 1833–1838, 1980.

JAPAN METEOROLOGICAL AGENCY, On the Izu-Oshima-Kinkai earthquake, 1978, *Rep. Coord. Comm. Earthq. Predict.*, **20**, 45–50, 1978 (in Japanese).

JOHNSTON, A. C., Localized compressional velocity decrease precursory to the Kalapana, Hawaii, earthquake, *Science*, **199**, 882–885, 1978.

JOHNSTON, M. J. S., Tectonomagnetic effects, *Earthq. Inf. Bull.*, **10**, 82–87, 1978a.

JOHNSTON, M. J. S., Tiltmeter studies in earthquake prediction, *Earthq. Inf. Bull.*, **10**, 182–186, 1978b.

JOHNSTON, M. J. S., Continuous strain measurements near the San Andreas fault, *Earthq. Inf. Bull.*, **10**, 187–191, 1978c.

JOHNSTON, M. J. S., B. E. SMITH, and R. MULLER, Tectonomagnetic experiments and observations in western U.S.A., *J. Geomag. Geoelectr.*, **28**, 85–97, 1976.

KANAMORI, H., Faulting of the great Kanto earthquake of 1923 as revealed by seismological data, *Bull. Earthq. Res. Inst., Univ. Tokyo*, **49**, 13–18, 1971a.

KANAMORI, H., Seismological evidence for a lithospheric normal faulting—The Sanriku earthquake of 1933, *Phys. Earth Planet. Inter.*, **4**, 289–300, 1971b.

KANAMORI, H., Tectonic implications of the 1944 Tonankai and the 1946 Nankaido earthquakes, *Phys. Earth Planet. Inter.*, **5**, 129–139, 1972a.

KANAMORI, H., Great earthquake and island arc, *Kagaku*, **42**, 203–211, 1972b (in Japanese).

KANAMORI, H. and S. MIYAMURA, Seismometrical reevaluation of the great Kanto earthquake of September 1, 1923, *Bull. Earthq. Res. Inst., Univ. Tokyo*, **48**, 115–125, 1970.

KELLEHER, J., Space-time seismicity of Alaska-Aleutian seismic zone, *J. Geophys. Res.*, **75**, 5745–5756, 1970.

KELLEHER, J., Rupture zones of large South American earthquakes and some predictions, *J. Geophys. Res.*, **77**, 2087–2103, 1972.

KELLEHER, J. and J. SAVINO, Distribution of seismicity before large strike slip and thrust-type earthquakes, *J. Geophys. Res.*, **80**, 260–271, 1975.

KELLEHER, J., L. SYKES, and J. OLIVER, Possible criteria for predicting earthquake locations and their application to major plate boundaries of the Pacific and Carribbean, *J. Geophys. Res.*, **78**, 2457–2585, 1973.

KELLEHER, J., J. SAVINO, H. ROWLETT, and W. MCCANN, Why and where great thrust earthquake occur along island arcs, *J. Geophys. Res.*, **79**, 4889–4899, 1974.

KERR, R. A., Prediction of huge Peruvian quakes quashed, *Science*, **211**, 808–809, 1981.

KING, C. Y., Radon emanation on San Andreas fault, *Nature*, **271**, 516–519, 1978.

KISSLINGER, C. and T. RIKITAKE, U.S.-Japan seminar on earthquake prediction and control, *EOS (Trans. Am. Geophys. Union)*, **55**, 9–15, 1974.

KOVACH, R. L., A. NUR, R. L. WESSON, and R. ROBINSON, Water level fluctuations and earthquakes on the San Andreas fault zone, *Geology*, **3**, 437–440, 1975.

KOYAMA, S. and Y. HONKURA, Observations of electric self-potential at Nakaizu (1), *Bull. Earthq. Res. Inst., Univ. Tokyo*, **53**, 939–942, 1978 (in Japanese).

KUNO, H., On the displacement of the Tanna fault since the Pleistocene, *Bull. Earthq. Res. Inst., Univ. Tokyo*, **14**, 619–631, 1936.

LATYNINA, L. A. and S. D. RIZAEVA, On tidal-strain variations before earthquakes, *Tectonophysics*, **31**, 121–127, 1976.

LEE, W. H. K., Earthquakes and China: A guide to some background materials, U.S. Geol. Surv., Open-file Rep., 88 pp., 1974.

LI, Quan-lin, Jin-biao CHEN, Lu YÜ, and Bai-lin HAO, Time and space scanning of the *b*-value—A method for monitoring the development of catastrophic earthquakes, *Acta Geophys. Sinica*, **21**, 101–125, 1978 (in Chinese).

LINDH, A., G. FUIS, and C. MANTIS, Seismic amplitude measurements suggest foreshocks have different focal mechanism than aftershocks, *Science*, **201**, 56–59, 1978.

LOMNITZ, C., Major earthquakes and tsunamis in Chile during the period 1535–1955,

378

Geol. Rundsch., **59**, 938–960, 1970.

MA, Hong-ching, Variations of the *b*-values before several large earthquakes occurred in North China, *Acta Geophys. Sinica*, **21**, 126–141, 1978.

MACCABE, M. P., Earthquake hazards reduction program 1978, U.S. Geol. Surv. Open-File Rep. 79–387, 113 pp., 1979.

MAP PRESS, *Province Maps of the People's Republic of China*, 169 pp., Map Press, Peking, 1971 (in Chinese written in Roman alphabets).

MARZA, V. I., The March 4, 1977 Vrancea earthquake seismic gap, *Bull. Seismol. Soc. Am.*, **69**, 289–291, 1979.

MATSUSHIMA, S., Variation of the elastic wave velocities of rocks in the process of deformation and fracture under high pressure, *Bull. Disas. Prev. Res. Inst., Kyoto Univ.*, **32**, 2–8, 1960.

MAZZELLA, A. and H. F. MORRISON, Electrical resistivity variations associated with earthquakes on the San Andreas fault, *Science*, **185**, 855–857, 1974.

MCCANN, W. R., S. P. NISHENKO, L. R. SYKES, and J. KRAUSE, Seismic gaps and plate tectonics: Seismic potential for major plate boundaries, in *Proceedings of Conference VI Methodology for Identifying Seismic Gaps*, J. F. Evernden (Convener), B. L. Isacks and G. Plafker (Organizers), pp. 441–584, U.S. Geol. Surv., Open-file Rep. 78–943, 1978.

MCGARR, A. and N. C. GAY, State of stress in the earth's crust, *Ann. Rev. Earth Planet. Sci.*, **6**, 405–436, 1978.

MIKUMO, T., M. KATO, H. DOI, Y. WADA, and T. TANAKA, Possibility of temporal variations in earth tidal strain amplitudes associated with major earthquakes, *J. Phys. Earth*, **25**, Suppl., S123–S136, 1977.

MJACHKIN, V. I., W. F. BRACE, G. A. SOBOLEV, and J. H. DIETERICH, Two models for earthquake forerunners, *Pure Appl. Geophys.*, **113**, 169–181, 1975.

MOGI, K., Migration of seismic activity, *Bull. Earthq. Res. Inst., Univ. Tokyo*, **46**, 53–74, 1968a.

MOGI, K., Sequential occurrences of recent great earthquakes, *J. Phys. Earth*, **16**, 30–36, 1968b.

MOGI, K., Rock breaking and earthquake prediction, *Zairyo (Materials)*, **23**, 320–331, 1974 (in Japanese).

MOGI, K., Rock mechanics and earthquake, in *Physics of Earthquake, Chikyu Kagaku*, 8, edited by H. Kanamori, pp. 211–262, Iwanami Shoten, Tokyo, 1978 (in Japanese).

MORRISON, H. F., R. FERNANDEZ, and R. F. CORWIN, Earth resistivity, self potential variations, and earthquakes: A negative result for $M=4.0$, *Geophys. Res. Lett.*, **6**, 139–142, 1979.

MORTENSEN, C. E. and M. J. S. JOHNSTON, The nature of surface tilt along 85 km of the San Andreas—Preliminary results from a 14 instrument array, *Pure Appl. Geophys.*, **113**, 237–249, 1975.

MUSHA, K., *Nihon Jishin Shiryo (Japanese Earthquake Records during 1848–1867)*, 1019 pp., Mainichi Shinbun Press., Tokyo, 1951 (in Japanese).

MUTO, K., A study of displacements of triangulation points, *Bull. Earthq. Res. Inst.*,

Univ. Tokyo, **10**, 384–391, 1932.

NATIONAL ACADEMY OF SCIENCES, *Earthquake Prediction and Public Policy*, 142 pp., National Academy of Sciences, Washington, D.C., 1975.

NATIONAL RESEARCH CENTER FOR DISASTER PREVENTION, Anomalously small value of the Ishimoto-Iida's coefficient m for foreshocks of Izu-Oshima Kinkai earthquake of January 14, 1978, *Rep. Coord. Comm. Earthq. Predict.*, **20**, 53–55, 1978 (in Japanese).

NATIONAL RESEARCH CENTER FOR DISASTER PREVENTION, Hydrofracturing stress measurements at Nishi-Izu Town, Shizuoka Prefecture, *Rep. Coord. Comm. Earthq. Predict.*, **22**, 104–107, 1977 (in Japanese).

NOAA, *Earthquake History of the United States*, 208 pp., Washington, D.C., 1973.

NORITOMI, K., Geoelectric and geomagnetic observations and phenomena associated with earthquake in China, in *Proceedings on the Chinese Earthquake Prediction by the 1977 Delegation of the Seismological Society of Japan*, pp. 57–87, Seismol. Soc. Japan, Tokyo, 1978a (in Japanese).

NORITOMI, K., Application of precursory geoelectric and geomagnetic phenomena to earthquake prediction in China, *Chin. Geophys.* (Am. Geophys. Union), **1**, 377–391, 1978b.

OHTA, H. and K. ABE, Responses to earthquake prediction in Kawasaki City, Japan in 1974, *J. Phys. Earth*, **25**, Suppl., S273–S282, 1977.

OHTAKE, M., Search for precursors of the 1974 Izu-Hanto-Oki earthquake, Japan, *Pure Appl. Geophys.*, **114**, 1083–1093, 1976.

OHTAKE, M. and M. KATSUMATA, Detection of premonitory change in seismic wave velocity, in *Symposium on Earthquake Prediction Research*, edited by Z. Suzuki, pp. 106–115, Sub-committee on Earthquake Prediction, National Committee for Geodesy and Geophysics, Science Council of Japan and Seismological Society, Japan, 1977 (in Japanese).

OHTAKE, M., T. MATUMOTO, and G. V. LATHAM, Seismicity gap near Oaxaca southern Mexico as a probable precursor to a large earthquake, *Pure Appl. Geophys.*, **115**, 375–385, 1977.

OHTAKE, M., T MATUMOTO, and G. V. LATHAM, Patterns of seismicity preceding earthquakes in Central America, Mexico and California, in *Proceedings of Conference VI Methodology for Identifying Seismic Gaps*, pp. 585–610, J. F. Evernden (Convener), B. L. Isacks and G. Plafker (Organizers), U.S. Geological Survey, Open-file Rep. 78–943, 1978.

OIKE, K., *Earthquake Prediction in China*, 235 pp., NHK Books, Nihon Hoso Kyokai, Tokyo, 1978a (in Japanese).

OIKE, K., Precursors and predictions of large earthquakes in China, in *Proceedings on the Chinese Earthquake Prediction by the 1977 Delegation of the Seismological Society of Japan*, pp. 135–148, Seismol. Soc. Japan, Tokyo, 1978b (in Japanese).

OIKE, K., Precursory phenomena and prediction of recent large earthquakes in China, *Chin. Geophys.* (Am. Geophys. Union), **1**, 179–199, 1978c.

OIKE, K., *China and Earthquake*, 261 pp., Toho Shoten, Tokyo, 1979 (in Japanese).

380

OIKE, K., R. SHICHI, and T. ASADA, Earthquake prediction in the People's Republic of China, *Zisin (J. Seismol. Soc. Japan)*, **28**, 75–94, 1975 (in Japanese).

PANEL ON EARTHQUAKE PREDICTION OF THE COMMITTEE ON SEISMOLOGY, *Predicting Earthquakes—A Scientific Technical Evaluation with Implications for Society*, 62 pp., National Academy of Sciences, Washington, D.C., 1976.

PRESS, F., H. BENIOFF, R. A. FROSCH, D. T. GRIGGS, J. HANDIN, R. E. HANSON, H. H. HESS, G. W. HOUSNER, W. H. MUNK, E. OROWAN, L. C. PAKISER, Jr., G. SUTTON, and D. TOCHER, *Earthquake Prediction: a Proposal for a Ten Year Program of Research*, 134 pp., Office Sci. Technol., Washington, D.C., 1965.

RALEIGH, B., G. BENNET, H. CRAIG, T. HANKS, P. MOLNAR, A. NUR, J. SAVAGE, C. SCHOLZ, R. TURNER, and F.WU, Prediction of the Haicheng earthquake, *EOS (Trans. Am. Geophys. Union)*, **58**, 236–272, 1977.

REDDY, I. K., R. J. PHILLIPS, J. H. WHITCOMB, D. M. COLE, and R. A. TAYLOR, Monitoring of time dependent electrical resistivity by magnetotellurics, *J. Geomag. Geoelectr.*, **28**, 165–178, 1976.

RESEARCH GROUP FOR ACTIVE FAULTS, *Active Faults in Japan*, 363 pp., University of Tokyo Press, Tokyo, 1980 (in Japanese).

RESEARCH GROUP FOR CRUSTAL STRESS IN WESTERN JAPAN, Absolute measurements of crustal stress by a stress relief method: (1) the Sazare Mine, Shikoku, *Rep. Coord. Comm. Earthq. Predict.*, **23**, 155–159, 1980 (in Japanese).

RIKITAKE, T., An approach to prediction of magnitude and occurrence time of earthquakes, *Tectonophysics*, **8**, 81–95, 1969.

RIKITAKE, T., Earthquake prediction research in the U.S.A., *Kagaku*, **44**, 571–574, 1974 (in Japanese).

RIKITAKE, T., Earthquake research in China—A summary of the U.S. seismological mission to China, *Kagaku*, **45**, 696–699, 1975a, (in Japanese).

RIKITAKE, T., Statistics of ultimate strain of the earth's crust and probability of earthquake occurrence, *Tectonophysics*, **26**, 1–21, 1975b.

RIKITAKE, T., Earthquake precursors, *Bull. Seismol. Soc. Am.*, **65**, 1133–1162, 1975c.

RIKITAKE, T., Dilatancy model and empirical formulas for an earthquake area, *Pure Appl. Geophys.*, **113**, 141–147, 1975d.

RIKITAKE, T., *Earthquake Prediction*, 357 pp., Elsevier, Amsterdam, 1976a.

RIKITAKE, T., Recurrence of great earthquakes at subduction zones, *Tectonophysics*, **35**, 335–362, 1976b.

RIKITAKE, T., Probability of a great earthquake to recur off the Pacific coast of Central Japan, *Tectonophysics*, **42**, 43–51, 1977a.

RIKITAKE, T., Classification of earthquake prediction information for practical use, *Rep. Coord. Comm. Earthq. Predict.*, **18**, 133–141, 1977b (in Japanese).

RIKITAKE, T., Earthquake prediction and warning, *Interdisciplinary Sci. Rev.*, **3**, 58–70, 1978a.

RIKITAKE, T., Biosystem behaviour as an earthquake precursor, *Tectonophysics*, **51**, 1–20, 1978b.

RIKITAKE, T., Classification of earthquake prediction information for practical use,

Tectonophysics, **46**, 175–185, 1978c.

RIKITAKE, T., *Can Animals Predict Earthquake?* 215 pp., Kodansha, Tokyo, 1978d (in Japanese).

RIKITAKE, T., Anomalous animal behavior preceding the 1978 Near Izu-Oshima Island earthquake, *Rep. Coord. Comm. Earthq. Predict.*, **20**, 67–76, 1979a (in Japanese).

RIKITAKE, T., Changes in the direction of magnetic vector of short-period geomagnetic variations before the 1972 Sitka, Alaska, earthquake, *J. Geomag. Geoelectr.*, **31**, 441–448, 1979b.

RIKITAKE, T., The Large-scale Earthquake Countermeasures Act and the Earthquake Prediction Council in Japan, *EOS (Trans. Am. Geophys. Union)*, **60**, 553–555, 1979c.

RIKITAKE, T., Classification of earthquake precursors, *Tectonophysics*, **54**, 293–309, 1979d.

RIKITAKE, T., Precursors to the 1978 near Izu-Oshima Island, Japan, earthquake of magnitude 7.0, in *Current Research in Earthquake Prediction I*, edited by T. Rikitake, Center for Academic Publications Japan/D. Reidel Publishing Company, Tokyo, 57–68, 1981a.

RIKITAKE, T., Anomalous animal behaviour preceding the 1978 earthquake of magnitude 7.0 that occurred near Izu-Oshima Island, Japan, in *Current Research in Earthquake Prediction I*, edited by T. Rikitake, Center for Academic Publications Japan/D. Reidel Publishing Company, Tokyo, 69–80, 1981b.

RIKITAKE, T., Earthquake precursors—A review, *Proceedings of the UNESCO Symposium on Earthquake Prediction*, 1982 (in press).

RIKITAKE, T. and M. SUZUKI, Anomalous animal behaviour preceding the 1978 Miyagi-ken Oki earthquake, *Rep. Coord. Comm. Earthq. Predict.*, **21**, 28–37, 1979 (in Japanese).

RIKITAKE, T. and Y. YAMAZAKI, Resistivity changes as a precursor of earthquake, *J. Geomag. Geoelectr.*, **28**, 497–505, 1976.

RIKITAKE, T. and Y. YAMAZAKI, Precursory and coseismic changes in ground resistivity, *J. Phys. Earth*, **25**, Suppl., S161–S173, 1977.

RIKITAKE, T. and Y. YAMAZAKI, A resistivity precursor of the 1974 Izu-Hanto-Oki earthquake, *J. Phys. Earth*, **27**, 1–6, 1979.

RIKITAKE, T. and I. YOKOYAMA, The anomalous behaviour of geomagnetic variations of short period in Japan and its relation to the subterranean structure, The 6th report (The results of further observations and some considerations concerning the influences of the sea on geomagnetic variations), *Bull. Earthq. Res. Inst., Tokyo Univ.*, **33**, 297–331, 1955.

RIKITAKE, T., A. TAKAGI, H. TAKAHASHI, S. UYEDA, H. AOKI, T. MATSUDA, Y. HAGIWARA, and M. KIMURA, Report on Earthquake Prediction in China—A Report of the 1978 Japanese seismological mission to China—, *Rev. Res. Disaster Prevention*, 44, 169 pp., National Research Center for Disaster Prevention, 1979 (in Japanese).

ROBSON, M., Role of the media in earthquake prediction, *Bull. N. Z. Natl. Soc. Earthq. Eng.*, **11**, 22–25, 1978.

RUNDLE, J. B. and M. MCNOTT, Southern California uplift—Is it or isn't it? *EOS (Trans. Am. Geophys. Union)*, **62**, 97–98, 1981.

SACKS, I. S., S. SUYEHIRO, D. W. EVERTSON, and Y. YAMAGISHI, Sacks-Evertson strainmeter, its installation in Japan and some preliminary results concerning strain steps, *Pap. Meteorol. Geophys.*, **22**, 195–208, 1971.

SANO, Y., Changes in the CA transfer functions at Kakioka, in *Proceedings of the Symposium on Conductivity Anomaly*, pp. 143–149, CA Group, Tokyo Inst. Technol., Tokyo, 1978 (in Japanese).

SASAI, Y. and Y. ISHIKAWA, Changes in the geomagnetic total force intensity associated with the anomalous crustal activity in the eastern part of the Izu Peninsula (2) —Izu-Oshima Kinkai earthquake of 1978—, *Bull. Earthq. Res. Inst., Univ. Tokyo*, **53**, 893–923, 1978 (in Japanese).

SATO, H., Some precursors prior to recent great earthquakes along the Nankai Trough, *J. Phys. Earth*, **25**, Suppl., S115–S121, 1977.

SATO, H. and N. INOUCHI, On relations between the ground uplifts and the earthquakes, in *Symposium on Earthquake Prediction Research*, edited by Z. Suzuki, pp. 138–144, Sub-committee on Earthquake Prediction, National Committee for Geodesy and Geophysics, Science Council of Japan and Seismological Society, Japan, 1977 (in Japanese).

SAVAGE, J. C. and R. O. BURFORD, Accumulation of tectonic strain in California, *Bull. Seismol. Soc. Am.*, **60**, 1887–1896, 1970.

SAVAGE, J. C. and R. O. BURFORD, Discussion of paper by C. H. Scholz and T. J. Fitch "Strain accumulation along the San Andreas fault," *J. Geophys. Res.*, **76**, 6469–6479, 1971.

SAVAGE, J. C. and W. H. PRESCOTT, Geodimeter measurements of strain during the southern California uplift, *J. Geophys. Res.*, **84**, 171–177, 1979.

SCHOLZ, C. H., A physical interpretation of the Haicheng earthquake prediction, *Nature*, **267**, 121–124, 1977.

SCHOLZ, C. H. and T. J. FITCH, Strain accumulation along the San Andreas fault, *J. Geophys. Res.*, **74**, 6649–6665, 1969.

SCHOLZ, C. H. and T. J. FITCH, Strain and creep in central California, *J. Geophys. Res.*, **75**, 4447–4453, 1970.

SCHOLZ, C. H., L. R. SYKES, and Y. P. AGGARWAL, Earthquake prediction: A physical basis, *Science*, **181**, 803–809, 1973.

SCIENCE AND TECHNOLOGY AGENCY, *"Binran" (handbook) of Earthquake Prediction*, 214 pp., Science and Technology Agency, Tokyo, 1977 (in Japanese).

SCIENCE AND TECHNOLOGY AGENCY, *An Interim Report on the System of Information about the Prediction of Tokai Earthquake*, 388 pp., Science and Technology Agency, Tokyo, 1978 (in Japanese).

SEISMOLOGICAL BUREAU OF ANHUI PROVINCE, *Macroscopic Anomaly and Earthquake*, 169 pp., Earthquake Press, Peking, 1978 (in Chinese).

SEISMOLOGICAL PARTY, BIOPHYSICAL INSTITUTE OF THE ACADEMIA SINICA, *Animal and Earthquake*, 99 pp., Earthquake Press, Peking, 1977 (in Chinese).

SEISMOLOGICAL SOCIETY OF JAPAN, *Proceedings of the Lectures by the Seismological Delegation of the People's Republic of China*, 85 pp., Seismol. Soc. Japan, Tokyo, 1976 (in Japanese).

SEISMOLOGICAL SOCIETY OF JAPAN, *Proceedings on the Chinese Earthquake Prediction by the 1977 Delegation of the Seismological Society of Japan*, 154 pp., Seismol. Soc. Japan, Tokyo, 1978 (in Japanese).

SEKIYA, H., The seismicity preceding earthquakes and its significance to earthquake prediction, *Zisin (J. Seismol. Soc. Japan)*, Ser. 2, **29**, 299–311, 1976 (in Japanese).

SEKIYA, H., Anomalous seismic activity and earthquake prediction, *J. Phys. Earth*, **25**, Suppl., S85–S93, 1977.

SEKIYA, H. and K. TOKUNAGA, On the seismicity near the sea of Enshu, *Rep. Coord. Comm. Earthq. Predict.*, **11**, 96–101, 1974 (in Japanese).

SEMYENOV, A. N., Variations in the travel time of transverse and longitudinal wave before violent earthquakes, *Izv. Acad. Sci. U.S.S.R. (Phys. Solid Earth)*, **4**, 245–248, 1969 (English edition).

SHIMAZAKI, K., Unusually low seismic activity in the focal region of the great Kanto earthquake of 1923, *Tectonophysics*, **11**, 305–312, 1971.

SHIMAZU, Y. and T. HIRAMATSU, Earthquake hazards and risk assessment of warning issue, *Zisin (J. Seismol. Soc. Japan)*, Ser. 2, **31**, 147–160, 1978 (in Japanese).

SHIZUOKA PREFECTURE, Report of the Shizuoka Prefecture Seismological Mission to China, 154 pp., Shizuoka Pref. Shizuoka, 1978 (in Japanese).

SIDORENKO, A. V., M. A. SADOVSKY, and I. L. NERSESOV, Soviet experience of prediction of earthquakes in the U.S.S.R. and the prospects for its development, Presented at the 1979 UNESCO Symposium on Earthquake Prediction, 1979 (in Russian).

SMITH, B. E. and M. J. S. JOHNSTON, A tectonomagnetic effect observed before a magnitude 5.2 earthquake near Hollister, California, *J. Geophys. Res.*, **81**, 3556–3560, 1976.

STUART, W. D., Diffusionless dilatancy model for earthquake precursors, *Geophys. Res. Lett.*, **1**, 261–264, 1974.

SUGISAKI, R., Geochemical prediction of earthquake, *Kagaku*, **48**, 704–710, 1978 (in Japanese).

SUGISAKI, R. and R. SHICHI, Precursory changes in He/Ar and N_2/Ar ratios of fault gases prior to earthquakes, *Zisin (J. Seismol. Soc. Japan)*, Ser. 2, **31**, 195–206, 1978 (in Japanese).

SULTANXODJAEV, A. N., I. G. CHERNOV, and T. ZAKIROV, Hydrogeoseismic precursors to the Gasli earthquake, *Rep. Acad. Sci. Uzbekistan*, **7**, 51–53, 1976.

SUZUKI, Z., Earthquake prediction research in China, *Gakujitsu Geppo* (Japan Society for the Promotion of Science), **31**, 523–527, 1978a (in Japanese).

SUZUKI, Z., General introduction to the China visit by the delegation of the Seismological Society of Japan in 1977, *Chin. Geophys.* (Am. Geophys. Union), **1**,

173–177, 1978b.

SUZUKI, Z., On the Chinese prediction of earthquakes, *Chin. Geophys.* (Am. Geophys. Union), **1**, 393–424, 1978c.

SUZUKI, Z., Procedure of earthquake prediction and alarm in the People's Republic of China, *Proceedings on the Chinese Earthquake Prediction by the 1977 Delegation of the Seismological Society of Japan*, pp. 7–32, Seismol. Soc. Japan, Tokyo, 1978d (in Japanese).

SYKES, L. R., Aftershock zones of great earthquakes, seismicity gaps, and earthquake prediction for Alaska and the Aleutians, *J. Geophys. Res.*, **76**, 8021–8041, 1971.

TANAKA, Y., Reports on observations of crustal stress, crustal deformation and gravity change, and their anomalous changes related to earthquakes in China, *Proceedings on the Chinese Earthquake Prediction by the 1977 Delegation of the Seismological Society of Japan*, pp. 89–111, Seismol. Soc. Japan, Tokyo, 1978a (in Japanese).

TANAKA, Y., Reports on observations of crustal stress and crustal deformation, and their anomalous changes related to earthquakes in China, *Chin. Geophys.* (Am. Geophys. Union), **1**, 425–442, 1978b.

TANG, Chi-yang, Bases for the prediction of the Lungling earthquake and the temporal and spatial characteristics of precursors, *Chin. Geophys.* (Am. Geophys. Union), **2**, 400–424; *Proceedings on the Chinese Earthquake Prediction by the 1977 Delegation of the Seismological Society of Japan*, pp. 13–32, Seismol. Soc. Japan, Tokyo, 1978 (in Japanese).

TERASHIMA, T., Change of V_P/V_S before the large earthquake of April 1, 1968 in Hyuganada, Japan, *Bull. Int. Inst. Seismol. Earthq. Eng.*, **12**, 17–29, 1974.

TSUBOI, C., Investigation on the deformation of the earth's crust found by precise geodetic means, *Japan. J Astron. Geophys.*, **10**, 93–248, 1933.

TSUBOI, C., K. WADATI, and T. HAGIWARA, Prediction of Earthquakes—Progress to date and plans for further development, *Rep. Earthquake Prediction Res. Group Japan*, 21 pp., Earthquake Research Institute, Univ. of Tokyo, Tokyo, 1962.

TSUBOKAWA, I., On relation between duration of precursory geophysical phenomena and duration of crustal movement before earthquake, *J. Geod. Soc. Japan*, **19**, 116–119, 1973 (in Japanese).

TURNER, R. H., J. M. NIGG, D. H. PAZ, and B. S. YOUNG, *Earthquake Threat—The Human Response in Southern California*, 152 pp., Inst. Social Sci. Res., Univ. California, Los Angeles, 1979.

USAMI, T., *Nihon Higai Jishin Soran* (Catalogue of Disastrous Earthquakes in Japan), 327 pp., Univ. of Tokyo Press, Tokyo, 1975 (in Japanese).

U.S. GEOLOGICAL SURVEY, Earthquake prediction— Opportunity to avert disaster, U.S. Geological Survey Circular 729, 35 pp., Washington, D.C., 1976.

UTSU, T., Large earthquakes near Hokkaido and the expectancy of the occurrence of a large earthquake off Nemuro, *Rep. Coord. Comm. Earthq. Predict.*, **7**, 7–13, 1972 (in Japanese).

UTSU, T., Detection of a domain of decreased P-velocity prior to an earthquake, *Zisin* (*J. Seismol. Soc. Japan*), Ser. 2, **28**, 435–448, 1975 (in Japanese).

UTSU, T., Probabilities in earthquake prediction, *Zisin (J. Seismol. Soc. Japan)*, Ser. 2, **30**, 179–185, 1977 (in Japanese).

UTSU, T., Calculation of the probability of success of an earthquake prediction (in the case of Izu-Oshima Kinkai earthquake of 1978), *Rep. Coord. Comm. Earthq. Predict.*, **21**, 164–166, 1979 (in Japanese).

UTSU, T. and A. SEKI, Relation between the area of aftershock region and the energy of the main shock, *Zisin (J. Seismol. Soc. Japan)*, Ser. 2, **7**, 233–240, 1955 (in Japanese).

WAKITA, H., Geochemical studies for earthquake prediction, in *Symposium on Earthquake Prediction Research*, edited by Z. Suzuki, pp. 165–175, Sub-committee on Earthquake Prediction, National Committee for Geodesy and Geophysics, Science Council of Japan and Seismological Society, Japan, 1977 (in Japanese).

WAKITA, H., Geochemistry related to earthquake prediction in China, *Proceedings on the Chinese Earthquake Prediction by the 1977 Delegation of the Seismological Society of Japan, pp. 113–134*, Seismol. Soc. Japan, Tokyo, 1978a (in Japanese).

WAKITA, H., Earthquake prediction and geochemical studies in China, *Chin. Geophys.* (Am. Geophys. Union), **1**, 443–457, 1978b.

WAKITA, H., Changes in level and chemical composition of underground water, in *Method of Earthquake Prediction*, edited by T. Asada, pp. 146–166, Univ. Tokyo Press, Tokyo, 1978c (in Japanese).

WALLACE, R. E., Goals, strategy and task of the earthquake hazard reduction program, U.S. Geol. Surv. Circ., No. 701, 26 pp., 1974.

WANG, M., M. Y. YANG, Y. L. HU, T. C. LI, Y. T. CHEN, Y. CHIN, and J. FENG, Mechanism of the reservoir impounding earthquakes at Hsinfengkiang and a preliminary endeavour to discuss their cause, *Eng. Geol.*, **10**, 331–351, 1976.

WARD, P. L., Earthquake prediction, *Rev. Geophys. Space Phys.*, **17**, 343–353, 1979.

WEISBECKER, L. R., W. C. STONEMAN, S. E. ACKERMAN, R. K. ARNOLD, P. M. HALTON, S. C. IVY, W. H. KAUTZ, C. A. KROLL, S. LEVY, R. B. MICKLEY, P. D. MILLER, C. T. RAINEY, and J. E. VAN ZANDT, *Earthquake Prediction, Uncertainty, and Policies for the Future*, 315 pp., Stanford Research Institute, Menlo Park, California, 1977.

WHITCOMB, J. H., Earthquake prediction-related research at the Seismological Laboratory, California Institute of Technology, 1974–1976, *J. Phys. Earth*, **25**, Suppl., S1–S11, 1977.

WHITHAM, K., M. J. BERRY, E. R. KANSEWICH, W. G. MILNE, and A. C. HEIDEBRECHT, *Report on the Canadian Seismological Mission to China, October 20–November 10, 1975*, 56 pp., Earth Phys. Branch, Dept. of Energy, Mines and Resources, Ottawa, Ont., 1975.

WILSON, J. T., Mao's almanac: 3,000 years of Killer earthquakes, *Saturday Rev. Sci.*, February 19, 60–64, 1972.

WORKING GROUP ON EARTHQUAKE HAZARDS REDUCTION, *Earthquake Hazards Reduction: Issues for an Implementation Plan*, 231 pp., Office of Science and Technology Policy, Executive Office of the President, Washington, D.C., 1978.

WU, F. T., Gas well pressure fluctuations and earthquakes, *Nature*, **257**, 661–663,

1975.

WYSS, M., A search for precursors to the Sitka, 1972, earthquake: sea level, magnetic field, and P-residuals, *Pure Appl. Geophys.*, **113**, 297–309, 1975.

WYSS, M., Local changes of sea level before large earthquakes in South America, *Bull. Seismol. Soc. Am.*, **66**, 903–914, 1976a.

WYSS, M., Local sea level changes before and after the Hyuganada, Japan, earthquakes of 1961 and 1968, *J. Geophys. Res.*, **81**, 5315–5321, 1976b.

WYSS, M., The appearance rate of premonitory uplift, *Bull. Seismol. Soc. Am.*, **67**, 1091–1098, 1977.

WYSS, M., Recent earthquake prediction research in the United States, in *Current Research in Earthquake Prediction I.*, edited by T. Rikitake, Center for Academic Publications Japan/D. Reidel Publishing Company, Tokyo, 81–127, 1981.

WYSS, M., R. E. HABERMANN, and A. C. JOHNSTON, Long term precursory seismicity fluctuations, in *Proceedings of Conference VI Methodology for Identifying Seismic Gaps*, pp. 869–894, J. F. Evernden (Convener), B. L. Isacks and G. Plafker (Organizers), U.S. Geological Survey, Open-file Rep. 78–943, 1978.

XU, Wen-yao, Kui QI, and Shi-ming WANG, On the short period geomagnetic variation anomaly of the eastern Kansu Province, *Acta Geophys. Sinica*, **21**, 218–224, 1978 (in Chinese).

YAMAGUCHI, R. and T. ODAKA, Precursory changes in water level at Funabara and Kakigi before the Izu-Oshima Kinkai earthquake of 1978, *Rep. Coord. Comm. Earthq. Predict.*, **20**, 60–62; *Bull. Earthq. Res. Inst., Univ. Tokyo*, **53**, 841–854, 1978 (in Japanese).

YAMAZAKI, Y., Precursory and coseismic resistivity changes, *Pure Appl. Geophys.*, **113**, 219–227, 1975.

YANAGIHARA, K., Secular variation of the electrical conductivity anomaly in the central part of Japan, *Mem. Kakioka Mag. Obs.*, **15**, 1–12, 1972.

YANAGIHARA, K. and T. NAGANO, The changes of transfer function in the Central Japan Anomaly with special reference to earthquake occurrences, *J. Geomag. Geoelectr.*, **28**, 157–163, 1976.

YOSHII, T., Changes in seismic wave velocities, in *Method of Earthquake Prediction*, edited by T. Asada, pp. 78–92, Univ. Tokyo Press, Tokyo, 1978 (in Japanese).

ZHAO, Yü-lin and Fu-ye QIAN, Electrical resistivity anomaly observed in and around the epicentral area prior to the Tangshan earthquake of 1976, *Acta Geophys. Sinica*, **21**, 181–190, 1978 (in Chinese).

ZHENG, Zhi-zhen, Yuan-zhuang LIU, Zuo-chun HU, Yuan-yuan LÜ, Tian-yong JING, and You-guo ZHANG, Investigation of preparatory process of Tangshan earthquake according to digital filtering of the data for underground water, *Acta Geophys. Sinica*, **22**, 267–280, 1979 (in Chinese).

ZHU, F. M., Prediction, warning and disaster prevention related to the Haicheng earthquake of magnitude 7.3, in *Proceedings of the Lectures by the Seismological Delegation of the People's Republic of China*, pp. 15–26, Seismol. Soc. Japan, 1976 (in Japanese).

SUBJECT INDEX

NAME INDEX